TOWARD MACH 2:
THE DOUGLAS D-558 PROGRAM

For sale by the U.S. Government Printing Office
Superintendent of Documents, Mail Stop: SSOP, Washington, DC 20402-9328
ISBN 0-16-049962-3

NASA SP-4222

TOWARD MACH 2: THE DOUGLAS D-558 PROGRAM

Edited by: J. D. Hunley

Featuring Comments by:

Stanley P. Butchart
Robert A. Champine
A. Scott Crossfield
John Griffith
Richard P. Hallion
and
Edward T. Schneider

The NASA History Series

National Aeronautics and Space Administration
NASA Office of Policy and Plans
NASA History Office
Washington, D.C.   1999

Library of Congress Cataloging-in-Publication Data

Toward Mach 2: The Douglas D-558 Program/edited by J.D. Hunley; featuring comments by Stanley P. Butchart . . . [et al.].
    p. cm.—(NASA history series)
    Papers of the NASA Dryden Flight Research Center Symposium on the D-558 Program, February 4, 1998.
    "NASA SP: 4222."
    Includes bibliographical references and index.
    1. High-speed aeronautics—United States—History—Congresses.
    2. Skystreak (Supersonic planes)—History—Congresses. I. Hunley, J.D., 1941- . II. Series.
TL551.5.T69 1999   99-11963
629.132'305'0973–dc21   CIP

Contents

Foreword xi

Introduction xii

Symposium Transcript

    Welcome by Edward T. Schneider 1
    Richard P. Hallion's Comments on the D-558-1 2
    Robert A. Champine's Recollections of the D-558-1 14
    John Griffith's Recollections of the D-558-1 20
    Richard P. Hallion's Comments on the D-558-2 28
    Stanley P. Butchart's Recollections of the D-558-2 and P2B 38
    A. Scott Crossfield's Recollections of the D-558-2 46
    Audience Questions and Panel Responses 56
    Wrap-up by Edward T. Schneider 60

Appendix — The Aircraft 61

Documents

1 Memo, Hartley A. Soulé, Discussion of D-558-1 airplane projects at NACA Headquarters on June 8, 1949, Date: June 13, 1949 64

2 Memo, Donald R. Bellman, Information concerning elevator vibration of the D-558-1 airplane, September 19, 1951 67

3 Memo, Donald R. Bellman, Progress report for the D-558-1 airplane (142) for the period September 22 to October 5, 1951, Date: October 12, 1951 73

4 Memo, Donald R. Bellman, Progress report for the D-558-1 (142) airplane for the period June 28 to July 11, 1952, Date: July 23, 1952 74

5 Memo, Donald R. Bellman, Progress report for the D-558-1 (142) research airplane for the period July 12 to July 25, 1952, Date: July 30, 1952 75

6 Memo, Donald R. Bellman, Progress report for the D-558-1 (142) research airplane for the period November 1 to December 1, 1952, Date: December 11, 1952 76

7 Memo, Herman O. Ankenbruck, Progress report for the D-558-II (37974) research airplane for the period August 25 to September 7, 1951, Date: September 13, 1951 77

8 Memo, Herman O. Ankenbruck, Progress report for the D-558-II (144) research airplane for the period September 22 to October 5, 1951, Date: October 16, 1951 78

9 Memo, Herman O. Ankenbruck, Progress report for the D-558-II (144) research airplane for the period November 3 to November 16, 1951, Date: November 21, 1951 81

10 Memo, Herman O. Ankenbruck, Progress report for the D-558-II (144) research airplane for the period December 1 to December 14, 1951, Date: December 18, 1951 82

11 Memo, Herman O. Ankenbruck, Progress report for the D-558-II (144) research airplane for the period June 28 to July 11, 1952, Date: July 23, 1952 .. 83

12 Memo, Herman O. Ankenbruck, Progress for the D-558-II research airplane (144) for the period November 1 to December 1, 1952, Date: December 11, 1952 84

13 Memo, Gareth H. Jordan, Progress report for the D-558-II (144) research airplane for the period September 1 to September 30, 1954, Date: October 6, 1954 86

14 Memo, Gareth H. Jordan, Progress report for the D-558 II (144) research airplane for the period September 1 to September 30, 1956, Date: October 3, 1956 87

15 Memo, Gareth H. Jordan, Progress report for the D-558-II (144) research airplane for the period November 1 to November 30, 1956, Date: December 4, 1956 88

16 Memo, Gareth H. Jordan, Progress report for the D-558-II (144) research airplane for the period December 1 to December 31, 1956, Date: January 8, 1957 89

17 Memo, James R. Peele, Progress report for the D-558-II (37975) research airplane for the period August 25 to September 7, 1951, Date: September 13, 1951 90

18 Memo, James R. Peele, Progress report for the D-558-II (37975) research airplane for the period September 8 to September 22, 1951, Date: September 28, 1951 91

19 Memo, Jack Fischel, Progress report for the D-558-II (145) research airplane for the period October 6 to October 19, 1951, Date: November 6, 1951 92

20 Memo, Jack Fischel, Progress report for the D-558-II (145) research airplane for the period October 20 to November 2, 1951, Date: November 8, 1951 94

21 Memo, Jack Fischel, Progress report for the D-558-II (145) research airplane for the period December 1 to 14, 1951, Date: December 18, 1951 96

22 Memo, Jack Fischel, Progress report for the D-558-II (145) research airplane for the period June 28 to July 11, 1952, Date: July 23, 1952 97

23 Memo, Jack Fischel, Progress report for the D-558-II (145) research airplane for the period July 26 to August 8, 1952, Date: August 15, 1952 98

24 Memo, Jack Fischel, Progress report for the D-558-II (145) research airplane for the period October 1 to November 1, 1952, Date: November 21, 1952 99

25 Memo, Jack Fischel, Progress report for the D-558-II (145) research airplane for the period November 1 to December 1, 1952, Date: December 11, 1952 101

26 Memo, Jack Fischel, Progress report for the D-558-II (145) research airplane for the period November 1 to December 30, 1954, Date: December 20, 1954 . 104

27 Memo, Jack Fischel, Progress report for the D-558-II (145) research airplane for the period September 1 to September 30, 1954, Date: October 6, 1954 105

28 Memo, Jack Fischel, Progress report for the D-558-II (145) research airplane for the period March 1 to 31, 1955, Date: April 18, 1955 106

29 Memo, Jack Fischel, Progress report for the D-558-II (145) research airplane for the period July 1 to July 31, 1955, Date: August 11, 1955 107

30 Memo, Jack Fischel, Progress report for the D-558-II (145) research airplane for the period June 1 to 30, 1956, Date: July 24, 1956 108

31 Letter, Ira H. Abbott, [NACA] Assistant Director for Research, To: Major General Frederick R. Dent, Jr., USAF, Commanding General, Wright Air Development Center, August 13, 1951 109

32 Letter, J.W. Crowley, [NACA] Associate Director for Research, To: Chief, Bureau of Aeronautics, Department of the Navy, Washington, D.C., Subject: Request for assignment of Navy omni[-]environment full-pressure pilot suit to NACA pilot of the D-558-II airplane, October 23, 1951 110

33 Letter, J.E. Sullivan, Director of Airborne Equipment Division, [Navy] Bureau of Aeronautics, To: National Advisory Committee for Aeronautics, Subject: Navy Omni-environment Full-pressure Suit, assignment of to the NACA pilot of the D-558-II airplane, November, 1951 112

34 Letter, F.A. Santner, Director, Naval Air Experimental Station, To: Chief, [Navy] Bureau of Aeronautics (AE-513), Subject: TED No. NAM AE 5101 Omni-environment full-pressure suit, research, development and test of; Flight test in the D-558-II airplane at Edwards Air Force Base, Edwards, Calif., 3 August to 4 September, 1953, Date: December 8, 1953 113

35 Letter, Walter C. Williams, Chief, [NACA] High-Speed Flight Research Station,

To: NACA, Subject: Choice of color for Research Aircraft at Edwards, December 3, 1951 116

36 Memorandum for Gordon S. Williams, News Bureau Manager, Boeing Airplane Company, April 25, 1952 [Subject: air-launch technique] 118

37 Letter, Walter C. Williams, Chief, NACA High-Speed Flight Research Station, To: NACA, Subject: Increased thrust of the LR8-RM-6 rocket engine, August 28, 1952 124

38 Letter, Walter C. Williams, Chief, NACA High-Speed Flight Research Station, To: NACA Headquarters; Attention: Mr. Clotaire Wood, Subject: Research Airplane Panel Meeting, January 2, 1953 126

39 Letter, Lieutenant Colonel Marion E. Carl, Senior School, Marine Corps Education Center, Marine Corps School, Quantico, Virginia, To: Chief, Bureau of Aeronautics, Aer-AC-241, Subject: Reports of flights in the NACA D-558-II, October 28, 1953 129

40 Memo, Donald R. Bellman, Aeronautical Research Scientist, To: NACA Headquarters, Subject: Rocket nozzle extensions used on the LR8 engine for the D-558-II, No. 144 airplane, January 25, 1954 136

41 Letter, Ira H. Abbott, [NACA] Assistant Director for Research, To: Chief, Bureau of Aeronautics, Department of the Navy, Subject: Results of use of the rocket nozzle extensions for the LR8 engine on the D-558-II airplane, February 3, 1954 139

42 Memo, Marion I. Kent, [NACA HSFRS] Administrative Officer, To: Mr. Bonney [NACA Headquarters], Subject: The research piloting experience of Joe Walker and Stan Butchart, April 29, 1954 141

43 Letter, De E. Beeler, Acting Chief, [NACA] High-Speed Flight Station, To: NACA Headquarters, Subject: Low temperature difficulties with hydrogen peroxide in the model D-558-II airplane, July 29, 1954 142

44 Letter, B.F. Coffman, Chief, Bureau of Aeronautics, To: National Advisory Committee for Aeronautics, Subject: Improved version of the LR8 liquid engine for use in the D-558-II airplane, June 2, 1954 143

45 Letter, Ira H. Abbott, [NACA] Assistant Director for Research, To: Chief, Bureau of Aeronautics; Department of the Navy, Subject: Improved version of the LR8 liquid rocket engine for use in the D-558-II airplane, July 23, 1954 146

46 Letter, B.F. Coffman, Bureau of Aeronautics, To: National Advisory Committee for Aeronautics, Subject: Improved version of the LR8 liquid rocket engine for use in the D-558-II airplane, August 11, 1954 147

47 Letter, John W. Crowley, [NACA] Associate Director for Research, To: Deputy Chief of Staff/Development, United States Air Force, Subject: Supply support for the B-29, NACA Serial No. 137, based at the NACA High-Speed Flight Station, Edwards, Calif., August 25, 1955 150

48 Letter, Joseph R. Vensel, Acting Chief, NACA High-Speed Flight Station, To: Chief, Bureau of Aeronautics, Subject: Completion of the D-558-II Research Program, June 17, 1957 151

Index 153

The NASA History Series 157

The four D-558 pilots with a model of the D-558-2 at NASA's Dryden Flight Research Center on February 4, 1998. From the viewer's left to right: Scott Crossfield, Stan Butchart, Bob Champine, and John Griffith. (NASA photo EC98-44406-2 by Tony Landis).

Foreword

In the long and proud history of flight research at what is now called the NASA Dryden Flight Research Center, the D-558 project holds a special place as being one of the earliest and most productive flight research efforts conducted here. Data from the D-558 and the early X-planes enabled researchers at what became NASA's Langley Research Center to correlate and correct test results from wind tunnels with actual flight values. Then, the combined results of flight and wind-tunnel testing enabled the U.S. aeronautical community to solve many of the problems that occur in the transonic speed range (about 0.8 to 1.2 times the speed of sound), such as pitch-up, buffeting, and other instabilities. This enabled reliable and routine flight of such aircraft as the century series of fighters (F-100, F-102, F-104, etc.) as well as all commercial transport aircraft from the mid-1950s to the present.

At the symposia honoring the 50th anniversary of the D-558-1 Skyrocket's first flight in February 1948, four D-558 pilots — Stanley P. Butchart, Robert A. Champine, A. Scott Crossfield, and John Griffith — plus Air Force Historian Richard Hallion offered insightful comments and meaningful anecdotes that deserved a wider audience than the few hundred people who attended. To make their recollections and related documents available to such an audience, NASA is publishing this volume. I am sure it will find a ready reception among the large group of people interested in the history of aviation.

Kevin L. Petersen
Director, Dryden Flight Research Center
February 1, 1999

Introduction

The Douglas D-558-1 Skystreak and D-558-2 Skyrocket were, with the Bell XS-1, the earliest transonic research aircraft built in this country to gather data so the aviation community could understand what was happening when aircraft approached the speed of sound (roughly 741 miles per hour at sea level in dry air at 32 degrees Fahrenheit). In the early 1940s, fighter (actually, in the terms of the time, pursuit) aircraft like the P-38 Lightning were approaching these speeds in dives and either could not get out of the dives before hitting the ground or were breaking apart from the effects of compressibility—increased density and disturbed airflow as the speed approached that of sound and created shock waves.

At this time, aerodynamicists lacked accurate wind-tunnel data for the speed range from roughly Mach 0.8 to 1.2 (respectively, 0.8 and 1.2 times the speed of sound, so named in honor of Austrian physicist Ernst Mach, who — already in the second half of the 19th century — had discussed the speed of a body moving through a gas and how it related to the speed of sound). To overcome the limited knowledge of what was happening at these transonic speeds, people in the aeronautics community — especially the National Advisory Committee for Aeronautics (NACA), the Army Air Forces (AAF — Air Force after 1947), and the Navy — agreed on the need for a research airplane with enough structural strength to withstand compressibility effects in this speed range. The AAF preferred a rocket-powered aircraft and funded the XS-1 (eXperimental Supersonic, later shortened to simply X), while the NACA and Navy preferred a more conservative design and pursued the D-558, with the NACA also supporting the X-1 research.

The flight research took place at the Muroc Army Air Field, with participation from a NACA contingent under Walter C. Williams that became the core of the later NASA Dryden Flight Research Center. While the D-558-1 with its jet engine was slower and less glamorous than the rocket-powered, air-launched XS-1, it flew for longer durations and thus gathered a lot of data more easily than its Bell counterpart. The D-558-2 was variously configured with jet and rocket engines, conventional takeoffs and air launchings. But the rocket-powered D-558-2 number 2 became the first aircraft to reach Mach 2.

The number 1 Skyrocket first flew on February 4, 1948. On the 50th anniversary of that date, the Dryden Flight Research Center held a symposium in honor of the event. It was introduced by current Dryden research pilot Edward T. Schneider and featured four of the original research pilots — Stanley P. Butchart, Robert A. Champine, A. Scott Crossfield, and John Griffith — talking about their experiences with the D-558 and its launch aircraft, the P2B-1S (Navy version of the B-29). In addition, Air Force historian Richard P. Hallion spoke about the Skystreak and the Skyrocket aircraft.

The previous night, the Center also held a symposium with a different format. Instead of each participant making a formal presentation, they all sat in a semicircle on stage and held a round-robin discussion, also with Ed Schneider as moderator.

Because all of the participants had valuable and interesting comments to make, it seemed imperative to preserve and print them so that those not privileged to attend the ceremonies could benefit from their recollections.

Naturally, there was a good bit of overlap in the information presented and stories told at the two sessions, so it would have been redundant to provide transcripts of both symposia. What I have chosen to do instead is to take as a basis the formal presentations made on the actual anniversary day and to integrate into them comments and anecdotes from the night before that were not included in the daytime session. Obviously, this violates the verbatim transcripts not only through the juxtaposition of related materials from two separate sessions, but also because I had to use my own words to create the appropriate transitions from one sentence or paragraph to another in the now-combined document. Despite such violation to the verbatim transcripts, I believe that the resultant narrative is true to the spirit of both sessions.

To ensure this, I have circulated the draft of this publication to the participants for their correction. I have also added footnotes to explain (or in a couple of instances, correct) comments made verbally from memory in front of an audience. The participants have contributed to the footnotes in a number of instances. In addition, I have appended historical documents from the National Archives about the D-558 program that add to the materials presented by the participants in the symposia. These are purposely scanned as documents into the study (rather than retyped) to give something of the flavor of looking at the documents themselves in an archive.

I believe the resultant publication adds significantly to the available literature on the D-558 flight research. It should be of interest to scholars, others interested in the history of aviation, and especially people working at or retired from the Dryden Flight Research Center. I would like to thank the participants in the symposia and Mrs. Gloria Champine for their help in getting their comments ready for publication. In addition, Tony Landis was very helpful in selecting photographs to illustrate the D-558 story and generously contributed some of these photographs from home to be scanned into the monograph. He, Peter Merlin, and Ed Schneider were kind enough to read the draft of this publication and offer corrections before it was sent to the participants. Besides Tony Landis, other members of the Dryden Photo Lab assisted in getting photographs assembled for this publication. I would be remiss, however, if I did not point out that I was not able to find several of the photographs used in the two symposia. Given the press of other projects competing for my time, I had to leave them out of this publication in the interest of getting it ready for printing. The All-Quality Secretarial Service of Morris Plains, New Jersey, professionally transcribed audio tapes from the two symposia, and Kelley Clark of OAO provided the tapes through the intermediacy of Lori Losey. Steve Lighthill did an artful job of laying out the typescript and photos, and Darlene Lister handled the copy editing in her usual professional way. I greatly appreciate the help all of these people provided.

J. D. Hunley, Historian
NASA Dryden Flight Research Center

NASA DRYDEN FLIGHT RESEARCH CENTER

SYMPOSIUM ON THE D-558 PROGRAM

INTRODUCTION: Ed Schneider

FIRST SPEAKER: Dr. Dick Hallion

DATE: February 4, 1998

PLACE: Dryden Flight Research Center

SCHNEIDER: Good morning, ladies and gentlemen. My name is Ed Schneider. Welcome to our presentation today.

Let me carry you back in time now to 50 years ago today, February 4, 1948. Here at Muroc, as it was known then, John F. Martin of the Douglas Company climbed into a D-558 Phase 2 Skyrocket, and lifted it off the ground for its very first flight. On November 20, 1953, Scott Crossfield flew another Skyrocket to a speed of Mach 2.005, to become the first man in history to fly faster than Mach 2.

Today is your chance to join us in a colloquium, which is going to be a piece of living history. Our very first speaker is Dr. Dick Hallion. And I would like to take some time now to introduce him. And from that point on, Dick will take you through the rest of the program.

Dick Hallion is the Historian for the United States Air Force in Washington, D.C., and directs its worldwide historical and museum programs. He's got a tremendous amount of experience in this area. Dick has a Ph.D. in aviation history from the University of Maryland and has been active as an author, and a curator, and a museum operator for many, many years. He's worked for the National Air and Space Museum. He has been the Chief Historian for the Air Force Flight Test Center. He worked in staff positions for the Aeronautical Systems Division at Wright Patterson Air Force Base. He was a visiting professor at the Army War College and then came back for a tour of duty with the Secretary of the Air Force. Since 1991, he's been the Air Force Historian in Washington.

Dick is a great friend of the DFRC [Dryden Flight Research Center]. He's a tremendous historian and communicator. He wrote a substantial portion of his book *Supersonic Flight*[1]—which, by the way, is on sale at the gift shop—at the age of 21 for his college thesis.

You know, one of the things that we're big on here at Dryden is our alliance with the Air Force Flight Test Center. And it's been very positive, especially under the leadership of our Director Ken Szalai, and the leadership of Air Force General Richard Engel. And both organizations take credit for many, many things. Well, you know, Dick's got a lot of time doing work for the Air Force. And I think we ought to be taking credit for him. So, for starters, I'm going to take credit for him as a NASA person today.

---

[1] Richard P. Hallion, *Supersonic Flight: Breaking the Sound Barrier and Beyond, the Story of the Bell X-1 and Douglas D-558* (rev. edn.: London and Washington, DC: Brassey's, 1997).

I could invest another 20 minutes going through a bio on Dick. If you want to read all the details about where he was born, where he went to school, and everything that he wrote — including 15 books — you can get that off the Internet. Dick, in short, is a recognized expert on research aircraft — as well, on the air war in the Persian Gulf. He's quoted frequently in air power magazines and treatises for use of air power in the present, and use of air power in future conflicts.

In fact, and this is a true story, his face has become so familiar that there is one executive producer of TV documentaries — I believe she's located on the east coast — who literally begged her staff not to bring in any more scripts with "Hallion" quoted as the expert. And the line she used was: "Doesn't anyone else in America qualify as an expert?" — or words to that effect. So, a true story. He really is an expert.

Some of the books that he's written — I know people have seen *Test Pilots: The Frontiersmen of Flight*[2] and the very famous *On the Frontier: Flight Research at Dryden, 1946-1981*.[3] Dick is going to set the context for our forum today. And he will take you through the rest of the afternoon, introducing our speakers. And now it's time to sit back before lunch and enjoy a piece of living history, "The Skyrocket D-558 Program — The X-Planes That Weren't." And we're going to learn why that is.

Welcome, Dick Hallion. [Audience applause]

HALLION: After an introduction like that, you can only go down. So it's with some foreboding that I approach the podium here.

It's a real pleasure to get back here. I've always enjoyed my personal association with Dryden. And I think it's very fitting today that we're here to commemorate what was an extraordinarily productive research aircraft program — the D-558 program, which historically is not necessarily as well appreciated as it should be. The D-558 aircraft were remarkable airplanes. They were intended originally for research in the transonic regime. And you had then one of the variants, the D-558-2, actually go out and make the first Mach 2 flight. How that occurred we'll be hearing about in some detail later from our panelists.

But first, let me discuss the context in which the D-558 program began. If you take a look at the history of aviation, you see that in mid-century we had two great revolutions. One of those was the turbojet revolution, which promised the ability to fly beyond 500 miles an hour. But at the very same time we had this promise, we had some very acute problems. We had some deficits in our aerodynamic knowledge, caused largely because of deficiencies in the state of wind-tunnel development and wind-tunnel testing. And so the second great revolution that comes along then is the flight research revolution, which basically is the reason why we have the whole transonic and supersonic flight breakthrough coming out of that.

This revolution has its origins as early as the 1920s really, when people start studying the phenomena of the airflow changes around propellers. And then it gets applied to fixed-wing aircraft in the 1930s. Because by the mid-1930s, we were starting to see accidents caused to experimental high-performance fighters. The first one which seems to have experienced this was the Messerschmitt Bf 109 in 1937,

---

[2] (rev. ed.; Washington, DC: Smithsonian Institution Press, 1988).

[3] (Washington, DC: NASA SP-4303, 1984).

which had an accident due to so-called compressibility effects.[4]

And then of course in this country very quickly we see this with the P-38, starting in 1941. And there's a tremendous acceleration of interest with World War II to try to close this transonic gap — this gap between Mach .75 and roughly Mach 1.2 — to find out what's taking place here. And although there were many different shortcut research methods developed, and although there was a tremendous stimulus for wind-tunnel research here, the real solution that people approached — largely within the National Advisory Committee for Aeronautics (NACA), and then within both the United States Army Air Forces and the United States Navy — was the idea to develop transonic and supersonic research airplanes. And out of that comes both the Army Air Force's program, which leads to the Bell XS-1, and then a Navy-sponsored program — the Douglas D-558.

Against this background, we have, early on, some tremendous national security needs. We're going from the World War II time period to a cold war time period. We recognize that there's a tremendous challenge to this country in terms of technological development, because we're locked head-to-head with the Soviet Union. And that's obviously going to be a very long confrontation. So there's a very strong desire and a very strong need here to master this whole field. And it's these aircraft that really contribute in a very great way to doing that.

Our first subject is the D-558-1 Skystreak. How did this program come about? There's a tremendous number of similarities in the way that the D-558 program as a whole came about, and the way the X-1 program came about. Both of them grew out of a need for a transonic research airplane. The National Advisory Committee for Aeronautics was very keen on developing some sort of aircraft, vaguely determined and relatively unspecified in terms of specific details, that could undertake transonic and low supersonic flight testing, and thereby address some of the problems that existed in the mid-1940s with the absence of available wind-tunnel technology to do reliable transonic testing.

There were two schools of thought. One of these favored a rocket-propelled airplane. That view was generally expressed by the Army Air Forces. And that climate of thinking resulted in the Bell XS-1. And the other school of thought favored higher-duration turbojet-powered aircraft. That was very much more in line with thinking expressed by NACA engineers, such as the legendary John Stack of Langley Laboratory. And out of this thinking came the D-558 program.

The two programs complemented each other extremely well. The XS-1 could reach high Mach numbers relatively quickly, of course, but had very little duration. The D-558 program could loiter, if you will, in the transonic regime, and collect a tremendous amount of data. What's very interesting in both cases is that there were key individuals in the services who played a major role in getting these programs going. For the Army Air Forces, Major Ezra Kotcher at Wright Field acted as the stimulus within the Army Air Forces to push this proposal. Within the United States Marine Corps, working for the Navy Bureau of Aeronautics, Lieutenant Abraham Hyatt drew up a specification for a transonic research airplane in late 1944.

And then also reflecting what happened in the X-1 program, you now had a requirement for a key industrial figure to become aware of what was going on, and to express corporate interest in developing such an aircraft. Well, in the case of the X-1,

---

[4] Increased density, a sharp rise in drag, and disturbed airflow at speeds approaching that of sound (Mach 1).

it was when Robert Woods, a Bell engineer, visited Wright Field in December of 1944, met with his old buddy Ezra Kotcher — and out of that came the X-1 program. And in the case of the D-558, it was an equivalent visit by a Douglas engineer named L. Eugene Root, who visited a buddy of his at the Bureau of Aeronautics named Commander Emerson Conlon — and with Conlon, became aware of the Hyatt specification. And as Root later said, he "grabbed it, and ran with it,"[5] and took it back to Douglas.

Now as those planes came along in early 1945, both committed to being straight-wing designs — different in configuration, but, nevertheless, straight-wing airplanes. There was a great deal of rising interest in the swept wing which, in this country, had been developed by Robert Jones at Langley, based on some work in industry that he had picked up on and elaborated from. And then, when we had the discovery in the rubble of Nazi Germany, of the Germans' tremendous interest in swept wings — which dated actually to the 1930s — that accelerated this kind of interest.

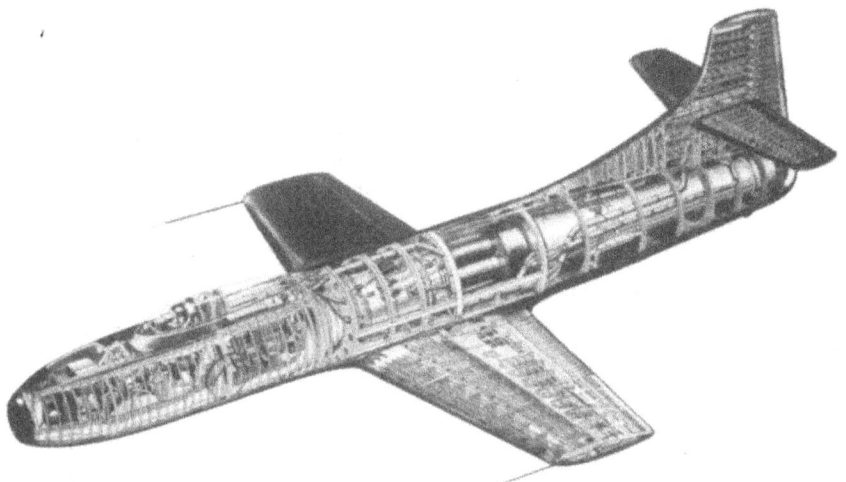

R. G. Smith watercolor showing cutaway view of the D-558-1 Skystreak (photograph provided by Tony Landis and reprinted with the permission of Boeing, of which the former Douglas Aircraft Corporation is now a part).

Both Bell and Douglas looked at swept-wing derivatives of their airplanes. In the case of Bell, they tried to put a swept wing in the X-1, decided it wouldn't work, and launched the X-2 program. In the case of Douglas, the firm simply had a slightly better situation. Its contract was for six airplanes. And the last three of those airplanes would have had differing wing configurations, in terms of thickness/chord ratio and aspect ratio.[6] And Douglas and the Navy got together and basically decided to take those last three airplanes and make them overtly swept-wing. And although you had this D-558 designation, in terms of actual design continuity between them,

---

[5] Quoted in Hallion, *Supersonic Flight*, p. 64, from a letter and recording Root sent to Hallion.

[6] The chord is the distance between the leading and trailing edges of an airfoil (the wing, in this case). Aspect ratio relates the span (distance from root to tip) of an airfoil to its area. A wing with high aspect ratio is long and slender; one with low aspect ratio is short and stubby.

The D-558-1 Skystreak under construction (photograph provided by Tony Landis).

they were really very different airplanes, as you can tell simply by looking at them. But the comparison would be the X-1 and the X-2 as basically representing that same philosophy for the Air Force-sponsored projects, and then of course the Navy-sponsored D-558-1 and -2 separately.[7]

So Douglas very quickly undertook design of a transonic research airplane, the idea being here to develop an aircraft that would use the sky as the laboratory. And one of the project engineers, A.M.O. Smith — project aerodynamicist — said their

---

[7] The three D-558-1 Skystreaks bore Bureau Numbers 37970 to 37972 and NACA "tail" numbers 140 to 142; the respective bureau and tail numbers for the three D-558-2 Skyrockets were 37973 to 37975 and 143 to 145.

The cockpit of the D-558-1 showing how it opened. (NASA photo E49-86).

The D-558-1 in flight in the late summer of 1947 (photo provided by Tony Landis).

task was basically to build the smallest airplane they could, wrapped around the largest airplane engine they could find. The painting by R.G. Smith [page 4], who was also a member of the Douglas design team and is now very well known as an aviation artist, really indicates to a very great degree what was meant by that.

You see here an aircraft that is indeed very tiny. It was quite cramped inside for the pilot. And it was literally packed with instrumentation. You had a wet wing to carry the fuel of the aircraft. You had specially designed, very thin wheels to retract within the wing. You could not use a conventional wheel arrangement. The wheels and tires had to be specially developed. And then, of course, you had the dominating feature, if you will — this very highly refined body-of-revolution type shape that indeed earned the airplane the nickname "flying test tube" — wrapped around this TG-180 engine.

The airplane was made primarily of aluminum, in terms of the wings and tail surfaces. The fuselage was constructed of aluminum framing, covered with magnesium sheeting. It carried 634 pounds of instruments, and had 400 pressure orifices buried within the wing — which was no mean feat in terms of building the wing at that time. The wing section was a NACA 65 section airfoil, 10 percent thickness:chord ratio. This tracked very much later with the number two XS-1 that was flown by the NACA, and which formed the basis for the X-1E, which is out here, of course, and which had itself a 10 percent thickness:chord ratio wing.[8]

The D-558 contract was an interesting contract. It specified six airplanes, for a total program price of $6,888,444.80. I don't know where they got the eighty cents. Now when you translate that into today's dollars, that's sixty-two million dollars which, for six research airplanes, I think we'd all agree is pretty much a bargain-basement price.

There was an intention to take the last three aircraft and to experiment with wings of varying aspect ratio and varying thickness:chord ratios. Ultimately, of course, those three aircraft were not built. Instead, that portion of the contract was set aside for the D-558-2. The original aircraft performance specification [for the D-558-1] was Mach .82 at sea level, corresponding to 625 miles per hour. And there was an 18 G ultimate load factor stipulated for the aircraft, which was the same load factor stipulated for the X-1.

There were two mock-up conferences on the airplane in July 1945 and August 1945. In August 1945 the program branched. And we got the substitution for the last three aircraft of a new swept-wing vehicle, the D-558-2. I will defer discussing the D-558-2 until this afternoon.

The first flight of the D-558-1 was on 14 April 1947 by Douglas test pilot Gene May. I have a photo here [page 5] that shows the aircraft under construction. I call your attention to the monocoque construction,[9] how the airplane came together. The number one airplane was the one that first flew in 1947. The number two airplane, which was the first NACA aircraft, was unfortunately the one in which Howard "Tick" Lilly died.

The number one airplane about the time of its first flight was a scarlet aircraft, nicknamed "the crimson test tube." You see how the cockpit opened on the aircraft.

Now the program moved very, very rapidly. In August 1947, flying both the number one and the number two airplanes, we had two official world airspeed records set in this aircraft. These broke a British record of 615 miles per hour that had been set earlier by Group Captain E.M. Donaldson in a Meteor.[10] The D-558-1 set initially a record of 640.663 miles per hour, flown by Commander Turner Caldwell. These were low altitude record runs, and then on 25 August 1947 — five days later — Marine Major Marion Carl reached 650.796 miles per hour.

Just as in June 1947 you had had a major research program outline developed for the X-1, split between the Army Air Forces and the NACA, in November 1947

---

[8] That is, the XS-1 number two had a ten percent thickness:chord ratio. The X-1E had a four percent thickness:chord ratio for its wings.

[9] A type of construction in which most of the stresses are carried by the covering or skin.

[10] Actually, as Hallion relates in *Supersonic Flight*, p. 141, there had been an intermediate record of 623.738 mph set by Army Air Forces Col. Albert Boyd in a P-80 on 19 June 1947.

Pilots Eugene F. May and Howard C. "Tick" Lilly (viewer's left to right) beside Douglas D-558-1 Skystreak number two, the one in which Lilly died. In this photograph, the Skystreak is painted bright red. (NASA photo E95-43116-8).

you had the same research directive come forth. Basically, Douglas would keep the number one airplane for its own purposes, and the NACA would get the number two and the number three airplanes.

In the latter part of the month, at the end of November 1947, we had the first NACA flight of the number two D-558-1 flown by Howard Lilly. Winter rains — which, of course, are no surprise given what we've had recently — winter rains closed the lakebed, and the plane did not resume flying until the following spring. Unfortunately on its nineteenth flight, on 3 May 1948, Lilly was killed when the compressor section of the TG-180 engine — the J35 engine[11] — disintegrated, severing flight control lines. The plane rolled inverted right after takeoff, and dove into the ground.

This caused Douglas to make extensive mods on the airplane, and indeed greatly influenced the subsequent history of the research aircraft program for the NACA in general — in that it put a great deal of emphasis upon ensuring that these research airplanes had such things as armoring of flight control systems that were designed to have significantly better safety characteristics than had been thought possible at that time. At the time of his death, Lilly was the first NACA pilot who had been killed in the line of duty.

In April 1949 we had the program resume, using the number three D-558-1. It was flown by Bob Champine, whom we're fortunate to have with us today. We can take a look at a couple of photos here. We have the classic red Skystreak shown here with Gene May. And despite that red color, it turned out that it was actually quite invisible at high altitudes. So there was a desire to repaint the airplane white to facilitate optical tracking. And, indeed, white became the standard color for the NACA research airplane fleet. Some portions — the flight control surfaces of the D-558-1 — were retained in red, the reason being that the flight control surfaces were extremely intolerant to changes in their overall weight and dynamic characteristics from having paint added to them. And they had to be left in red, lest there be the possibility for flight control surface flutter problems.

---

[11] The Allison J35-A-11 had originally been developed by General Electric as the TG-180.

This head-on shot shows how you had a bifurcation in the inlet. If you take a look at the inlet, obviously it splits then and goes around the pilot. So that even though you have what looks like a nice roomy circle of cross-section fuselage — the actual little capsule, if you will, that the pilot fits in is really quite narrow.

Head-on view of the D-558-1 showing the bifurcation in the engine inlet, forcing the intake air to go on either side of the pilot. (NASA photo E49-89).

The airplane had some interesting construction approaches for its time. It had aluminum framing for the fuselage, covered with magnesium sheeting. And then it had aluminum wing and tail surfaces. And even more interestingly, you had those 400 orifices cut into that wing for pressure distribution measurements. When you think about the standards of construction for that airplane at the time, it was really kind of a tribute to Ed Heinemann's design team[12] that it was able to do that as well as it did. It was really extraordinary.

To increase mission endurance, the plane was flown with tip tanks. And we have here just sort of "the sweet nostalgia of the never-to-be-forgotten moment." We have a nice little photograph here [page 10] showing the airplane in its classic NACA markings in white — the number three D-558-1 cruising right along on one of its transonic research missions.

What did the test flights show on this program? Basically, the D-558-1 configuration exhibited a marked increase in wheel force for trim, as Mach number went from about 0.82 to 0.87. It went from about five pounds push to about 30 pounds push [depending upon the incidence of the movable horizontal stabilizer[13]]. Lateral

---

[12] Edward H. Heinemann was the chief engineer with the Douglas Aircraft Corporation who headed the design team for the D-558-1 and D-558-2. See Hallion, *Supersonic Flight,* esp. pp. 63-5, 167.

[13] See Melvin Sadoff, William S. Roden, and John M. Eggleston, *Flight Investigation of the Longitudinal Stability and Control Characteristics of the Douglas D-558-1 Airplane (BUAERO No. 37972) at Mach Numbers up to 0.89* (Washington, D.C.: NACA Research Memorandum L51D18, 1951), esp. pp. 1, 6-7, 18. Thanks to Ed Saltzman for calling this source to my attention and interpreting it.

stability deteriorated over the same speed range, and there was pronounced wing dropping experience with this aircraft above Mach 0.84.

The NACA was very interested in this and, as a result, undertook many studies here of lateral, longitudinal, and dynamic stability characteristics of the aircraft. One of the most distinctive things added to the airplane was a series of vortex generators, for which Boeing should forever give thanks, employed to stabilize air flow. They worked very well. They were adapted subsequently for a whole range of aircraft — the B-47, the B-52 which had great rows of them, the KC-135, on to the 707 family, up to the 757, 767, and 777 of the present day and, for that matter, the Douglas A-4 as well. And this became sort of a hallmark, as a quick fix of the early supersonic and transonic era.

There was an extensive longitudinal stability research program flown with the D-558-1 number three airplane in 1950 and 1951. And then that was followed in

D-558-1 in flight on one of its transonic research missions. (NASA photo E-713).

1952 by an equally extensive lateral stability investigation. The longitudinal stability program consisted primarily of abrupt pull-ups. The lateral stability program consisted of taking data during abrupt rolls. And then there was a brief dynamic stability program undertaken in the program in 1953, consisting primarily of elevator and rudder pulses before the aircraft was relegated basically to use as a test pilot trainer. It finally made its last flight on 10 June 1953. And it was returned to the United States Navy in dead storage in 1954.

It's very interesting to take a look at the D-558-1, as distinct from the -2 airplane, because it was playing Avis, if you will, to the X-1's Hertz. But at the same time, this airplane — as Scott Crossfield and others have pointed out — was absolutely critical to giving us a thorough understanding of what was happening in the transonic regime. It's also very interesting to me, taking a look at both programs — the -1 and the -2 — to compare the tremendous success we had with these relatively complex aircraft and to contrast that with what was happening in Europe at the same time where you had, indeed, a whole series of false starts, dashed hopes, dashed expectations of whole families of research airplanes that were being developed in Great Britain, in France, and elsewhere — where there were tremendous national resources and industrial resources going into these. And the programs were not going along at anywhere near the pace that they should.

We'll certainly hear a lot more from our guests this afternoon when we resume the conference. But I would hold that one of the key things in making the American program a success was not merely the design of the airplanes — because the airplanes were very well thought out and extremely well crafted — which, as I've said, is a tremendous tribute to the design team headed by Ed Heinemann at Douglas, but I think also a tremendous tribute to the NACA here at Muroc, which was then the High-Speed Flight Research Station headed by Walt Williams.

And Walt Williams, of course, is a name that's not unfamiliar, certainly, to people in this audience.[14] But it's well worth mentioning, again, that in my view, Walt Williams was probably the finest flight test researcher and research director that this country produced. His impact and his imprint was on every major aerospace revolution, literally going from the transonic era of the late 1940s all the way through the landing on the moon in 1969 and beyond.

We have a number of people who will be receiving due mention and deserved mention today, and I would think that it's very fitting that the first of those that we single out for special mention is the late Walter C. Williams who, of course, loved this Center with the same intensity and passion that he brought to the love of aviation in general.

I would also like to point out that the very fine audiovisual materials which you'll be seeing today have been pulled together, particularly in the case of the photos that I'm using, from the photo archives here at Dryden, which is a unique historical resource. And we have Tony Landis to thank for that. And I appreciate Tony's doing that very much.

And I would also like to mention that in addition to the very distinguished guests we have here today who flew the aircraft and who maintained the aircraft, we have a member, indeed, from the Douglas design team — who worked on this aircraft — Charlie Delavan. And Charlie, if you'd stand up — I'd like you to take a bow. Because [audience applause] without people like Charlie, we certainly would not be able to have this symposium we're having today.

[BREAK FOR LUNCH]

HALLION: We now come, I think, to the real meat of the program. We're going to have presentations by our very distinguished panel of guests. First, two presentations on the D 558-1. Then we'll have a discussion on the D-558-2, followed by some presentations on the D-558-2. So what I'd like to do at this point is to have our distinguished guests please stand up. Bob Champine and his lovely wife, Gloria; John Griffith and his wife, Maxine; Stan Butchart; and Scott Crossfield. We're

---

[14] Williams was the first head of what later became the Dryden Flight Research Center, where he was instrumental in the successes of the early research aircraft and helped prepare the X-15 program before leaving the High-Speed Flight Station in September 1959 to become Associate Director of NASA's newly formed Space Task Group created to carry out Project Mercury. After serving with the Aerospace Corp. on the Gemini and Titan III vehicles, he joined NASA Headquarters as Chief Engineer.

honored, indeed, gentlemen that you're here today. I also want to recognize in the audience Donna Termeer, who's here from Assemblyman George Runner's office. Donna, welcome to the session. We're delighted to have you with us today.

If we take a look at visual images of the D-558-1 and D-558-2 [scattered throughout this volume], we see how evocative these aircraft were, and you think that there's a revolution taking place in aviation at this time. We're seeing a radical transformation literally, from the era of propeller-driven airplane to the era of the supersonic jet aircraft. They had a certain beauty, I think, that was all their own, and frankly, the shapes were extremely evocative. I think when you look at something like an F-86 or the D-558-2, that we haven't developed any aircraft since that time that really had that same degree of elegance. There was something there that I think resonates very deeply with us.

Walt Williams, Scott Crossfield, and Joe Vensel, Flight Operations Manager, (viewer's left to right) beside the D-558-2 on November 20, 1953, the day Crossfield exceeded Mach 2. (NASA photo E-1097).

As I said earlier, we're very honored to have the individuals who actually flew these aircraft with us today. We'll start first with recognition of two individuals who played a major role in the D-558 program — Bob Champine and John Griffith. These individuals — both of them — had very distinguished flying careers.

Bob Champine graduated from the University of Minnesota in 1943 with a bachelor's degree in aeronautical engineering, went through the civilian pilot training program, and became a naval aviator. After leaving active duty in 1947, he joined the staff of the NACA's Langley Memorial Aeronautical Laboratory at Hampton, Virginia. He did a lot of work there on an airplane that's a relative rarity — people don't think of it too much — the Bell L-39, which was a swept-wing variant of the P-63 Kingcobra. Despite that "39," it had no relationship to the P-39. It was a very important low-speed, swept-wing test bed, for a number of swept-wing aircraft, including the F-86, the D-558-2, and the Bell X-2. He was transferred out to Edwards in October 1948, did early research flying on both the X-1 and the D-558 program, went back to Langley, did a tremendous amount of work at Langley through the years on a whole range of aircraft, from high performance airplanes through vertical take-off and landing aircraft. Bob became Langley's Senior Test Pilot and retired in January 1979, two days after making his last research flight at NASA's Wallops facility in a CH-47 helicopter. A 31-year career. An utterly distinguished gentleman.

The other individual who's here today is equally distinguished — John Griffith. John undertook some studies at Thornton Township Junior College in Harvey, Illinois, graduating as valedictorian in pre-engineering. He went into the Army Air Forces in November 1941, served in the war in the South Pacific, and flew 189 missions in New Guinea in some very tough times, under some very daunting conditions. He was awarded two Distinguished Flying Crosses and four air medals for service in New Guinea. He left the service in October 1946, went back to study aeronautical engineering at Purdue University, and graduated with honors in aeronautical engineering from that university. He then joined the NACA at Cleveland, where he did some very interesting work in early ramjet testing. That was one of Cleveland's big projects in those days. Some icing research work — something else they were very well known for.[15] Then, of course, he came out here in August of 1949 and flew in the early X-series aircraft — the X-1, the X-4, the D-558 program. He left the NACA in 1950, joined Chance Vought, and worked there for a period of time doing experimental flight tests on the F7U Cutlass, had a career with United Airlines, with Westinghouse as the Chief Engineering Test Pilot, and a six-year career with the FAA doing a lot of work assisting in the attempt to develop the first supersonic transport. He had a second tour of duty as a flight instructor with United and flew the line with them for about seven years. We're very fortunate to have John here as well.[16]

These individuals — and certainly when we talk about Stan and Scott later — you'll see that these were very tough individuals. They were tough individuals dealing with very difficult times. They did very well, and have continued to do so.

To give you an example, Bob here — hale and hearty as he looks — Bob is recovering from a stroke — fortunately mild. He had it two months ago. He's made a remarkable comeback. As a result, Bob doesn't feel terribly comfortable at times speaking. And so, Bob, I'd like you to stand and once again be recognized by our audience. He's left a very fine written memoir that he prepared for this conference. But he's discussed this with John, and John will be handling Bob's portion of the discussion here on the D-558.

So at this point, Bob, I'd like you and John to stand up. And, John, you can

---

[15] The NACA's laboratory in Cleveland, established in 1941 and renamed in honor of George W. Lewis, NACA Director of Aeronautical Research from 1924 to 1947, in 1948, participated substantially in the NACA's studies of aircraft icing in this period.

[16] These introductions include elements from Ed Schneider's introductions the previous night and additions by John Griffith.

come forward to the podium. [Audience applause]

GRIFFITH: This is a paper that Bob and his wife have prepared. And I will read it as written here:

Good afternoon. I'm happy to be with you today. I thank Mr. Kenneth Szalai, Center Director, and Mr. Cam Martin of External Affairs for inviting us to Dryden, particularly to be with my fellow aviators John Griffith, Scott Crossfield, Stan Butchart, Ed Schneider, and our good friend Dr. Richard Hallion, without whose dedicated research effort and pilot interviews the detailed history of supersonic flight would be forever lost. The complete records just do not exist today.

I'm Bob Champine — a kid who grew up in Minneapolis, Minnesota, with my eyes to the heaven and my heart with wings. I used to ride my bicycle over to World Chamberlin Airport in Minneapolis to clean out hangars, wash airplanes, and do whatever I could to be offered just one ride — a ride in an airplane. My first ride was in a Fleet. I was about 12, and didn't tell my mother because she might not let me go to the airport again. I started flying a Piper Cub in the summer of 1939 and soloed in July 1940. I had to do a lot of sweeping, washing, and polishing airplanes to get that far. Between my building model airplanes, and competing in model meets — winning a lot of the time — and time I spent at the airport, I didn't have a lot of time for my school studies. I was just an average student.

When I graduated from Roosevelt High School in 1939, I wanted to learn to fly. But my mother said, "If you like airplanes so much, you need to learn to design them. I want you to go to college and study to be an aeronautical engineer." This was difficult because of my less-than-exceptional grades. After several conferences with my high school principal and also with the admissions office, I was finally accepted at the University of Minnesota in the Institute of Technology's Aeronautical Engineering Program, backed both by my mom and stepfather, Clifford Champine, who agreed to pay my tuition. I started college that fall. I really had to buckle down and study, as studying did not come easy to me. It was difficult. But my drive to become an aeronautical engineer made the difference.

While I was in college, World War II started, and I began primary flight training under the Naval Civilian Pilot Training Program, and upon graduation in 1943, was commissioned an ensign in the U.S. Navy. Since I wanted to be a naval aviator, I had to give up my commission and enroll in the Naval Cadet Program at Pensacola. At the end of my training, I was commissioned a naval aviator, and my mother pinned my wings on me in Pensacola. That was a proud day.

In 1947 when my Navy term was up, I was stationed at the Naval Air Base in Norfolk, Virginia, flying [F4U] Corsairs. Through my studies at the University of Minnesota, I learned of the outstanding reputation of the National Advisory Committee for Aeronautics, which was just across the river from Norfolk at Hampton, Virginia. With the approval of my superior in Norfolk, I flew my Corsair over to NACA, landed in front of the hangar, and rolled up to the large office building attached to the hangar. I swung the tail around smartly, folded up the Corsair's wings, climbed out, and asked, "Who's the boss here? I would like to talk to him." Of course, everyone there was looking out their windows watching me, and I had no trouble locating the head of the Division, Mr. Mel Gough, and head of the pilots, Mr. Herbert Hoover. I told them I was coming out of the Navy shortly, and would like to come to work for NACA Langley Laboratory as a test pilot. I was told to fill out the government forms for employment and they would look at them. The requirement for a test pilot was 1,000 hours of single-engine and transport flying, and I had only a little over 900.

Well, they offered me a job as an aeronautical engineer scientist, and I wasn't too happy about that. I wanted to be a test pilot, and not a scientist behind a desk. I told them that if they could not hire me as a research pilot, then I was going to use my G.I. Bill and go to helicopter school at Sikorsky in Connecticut. After discussions with Mr. Hoover, Mr. Gough said, "Aw, hell, come on with us as a research pilot and we'll teach you to fly helicopters here at Langley." I accepted.

After receiving my discharge from the Navy, I remained with the Naval Reserve. I found a room in a home in Hampton, Virginia, and began working at NACA in December of 1947. Was I thrilled! Langley had many airplanes and helicopters, and I was just itching to get my hands on the controls. Not long after I was hired, Mr.

Bob Champine and Herb Hoover beside the XS-1. (NASA photo E49-5).

Herbert Hoover, my mentor and dear friend, gave me a manual and told me to take it home; we were flying the B-29 in the morning. This is the way a lot of my training went: read the manual, and then we would go out flying.

I was just thrilled with the opportunity I had and didn't realize that my salary was only about $50.00 a week. I had saved $5,000 in the Navy and bought an old Ford car, and I was just the happiest soul on earth. Everyone in the Pilot's Office knew about a super-secret project that was going on in the California desert at the time. But I was happy where I was with just the best job in the world. I was a test pilot with NACA. It doesn't come any better than that.

In California, the XS-1 aerodynamic research program was continuing. Chuck Yeager broke the sound barrier in the Air Force XS-1 6062 October 14, 1947, and Herb Hoover, the second man, broke the sound barrier on March 10, 1948, in the NACA XS-1 6063.[17] Howard Lilly from the NACA Cleveland Laboratory and

---

[17] The full designations for these two aircraft were 46-062 and 46-063, but they bore the shortened designations 6062 and 6063 on their tails.

Hoover continued with the research program, and on March 31, 1948, Lilly exceeded Mach 1 on his third XS-1 flight. In May 1948 he was killed on his nineteenth flight of the D-558-1 number two.

Hoover needed another pilot at Muroc, and quietly approached me about going out there. I didn't know at the time that other pilots had been approached and, for various reasons, turned the assignment down. I was thrilled to say yes, but I had two conditions: (1) Let me fly all the planes Langley had before I went to California, and (2) I would return to my job at Langley. I had a ball flying everything in the hangar at Langley and being under the wing of Herbert Hoover who, behind closed doors at Langley, gave me critical instruction on the flying qualities of the XS-1 number two, and on NACA's aerodynamic research program.

When he decided I was ready, I left Langley and drove my old Ford out to Muroc in October 1948. Hoover remained at Muroc to train me, and on November 1, 1948, he turned the X-1 over to me. I made my first flight on November 23rd. I became the sixth [pilot in the XS-1 and D-558 series] to reach Mach 1 December 2, 1948, on my fourth flight.

This table shows research flights that John Griffith and I flew in 1948 through 1950:

Aircraft	Bob Champine NACA Research Pilot	John Griffith NACA Research Pilot
X-1	13 flights	9 flights
X-4	0 flights	3 flights
D-558-1	9 flights	16 flights
D-558-2	12 flights	8 flights

During my X-1 flights, there were a couple of incidents I would like to share with you. After settling in the X-1 beneath the B-29, I experienced a radio failure prior to launch. Using my knee pad, I wrote a note on a flight card "secure the drop," which, in my Navy lingo, meant stop. The note was passed through the bomb bay to the Air Force crew on the B-29. They thought everything was secure. They dropped me! I had to scramble to get the radio working. But I had it fixed and completed the flight okay.

On another flight, the cockpit camera just over my shoulder broke loose during the flight, and went slamming around inside the cockpit. I began to jerk the wires out and stash the camera beneath my leg, but not before it cracked the inner windshield. As I was attempting to land, the windshield frosted over, and I could not see. I put my thumb on the windshield, and melted a very small spot. I was able to put my eye close to it and see well enough to land on the dry lakebed. [Aside by John Griffith:] I might say that any of you who have seen the X-1 could see that the visibility for landing was not the best, since they wanted the windscreen to conform to the shape of a .50 caliber rifle bullet. But sometimes it did frost over, and then the chase plane would be telling you how high you were, and you hoped that you'd hit the lake at the right attitude. If you hit the lake with the nose wheel first, the X-1 was out of control, and there were a lot of people, including myself, who started bouncing along the lake as a result of the nose wheel hitting first and breaking off. Anyway, to get back to Bob's story here:

This photo gives you an aerial view of the base, with Rogers Dry Lake in the center. When we were here, there was a railroad going across. But we still had seven miles north and south and five miles east and west. Usually when we were landing the X-1, we'd shoot for about the one-mile marker, and usually didn't miss it by very

Aerial view of what was then (1948-49) called Muroc Air Force Base (now Edwards AFB) and vicinity. In the center, shaped somewhat like an hourglass, is Rogers Dry Lake (sometimes referred to as Muroc Dry Lake). (Photograph supplied by Bob Champine, available as NASA photo EC98-44613-1).

much. [Comment by Griffith:] I think there was one day that Bob was a little low on the base leg and he said, "Please advise." And the only advice I could think of was, "Our Father, who art in heaven."

Here is a picture of myself, Chuck Yeager, and Herb Hoover. The next is a picture of Mr. Hoover when he received the Air Force Association Award in 1948 for

Bob Champine, Chuck Yeager, and Herb Hoover (viewer's left to right) standing next to an X-1. (NASA photo EC98-44613-4, originally supplied by Bob Champine).

Herb Hoover with his Air Force Association Award in 1948. (NASA photo EC98-44613-7, originally supplied by Bob Champine).

his flight as the first civilian and the second man to break the sound barrier. He also received the Octave Chanute Award that year.

In the next photo, you can see just how small the X-1 was compared to the B-29 drop plane. Another one lets you see how we entered and exited the aircraft. Once inside, we were in. No thought of escape; we had to land it. Research flights were of short duration — maybe about 15 minutes of actual flight time. Then days and sometimes weeks would pass before there was another flight, and I was anxious for more flight time. I made good use of my Naval Reserve status, and was assigned my weekend warrior duty at Los Alamitos, California. Since I didn't want anyone to know my actual job at Muroc, I would take the Muroc C-47 and fly it by myself, would park about a mile away, and walk down the flight line to my assignment as an ensign in the Naval Reserve. I was able to get a lot of flying time there and had great fun.

X-1 predominantly flown by NACA pilots next to its B-29 "mothership." (NASA photo E-9).

Bob Champine exiting the X-1. (NASA photo EC98-44613-2, originally supplied by Bob Champine).

The next photograph shows our X-1 (6063), after it was modified for other research flights, as it stands today — proudly, in front of the Administration Building here at Dryden.

Original NACA X-1, modified as the X-1E, in front of the Headquarters Building at NASA Dryden. (NASA photo ECN 12506).

During the X-1 period, NACA took delivery of the D-558-1 number one Skystreak which was relegated to spares support. NACA test pilot Howard Lilly flew the D-558-1 Skystreak on its first NACA flight in November 1947, about six months before he was killed after engine failure on takeoff on May 3, 1948. In 1948, the D-558-1 Skystreak number three was delivered to NACA, and I made NACA flight one on April 22, 1949, for pilot familiarization.

The next two photos show the Skystreak on the ground, and then in flight. I made nine flights in the D-558-1 Skystreak and 12 flights in the D-558-2 Skyrocket, making NACA flight one on May 24, 1949. The next photo [page 22] shows the Skyrocket on the ground, taking off with a JATO assist.[18]

My thirty-two years as a test pilot for NACA/NASA were wonderful times — from flying the X-1, to spacecraft rendezvous, and simulated landings on the moon. I had it all. Thank you. [Audience applause]

GRIFFITH: Well, to turn to my own experiences, as Dick said, I grew up in Homewood, Illinois, near Chicago. And Green Three went very close to our house, which was one of the early air routes that went from Chicago, to Goshen, to Toledo, to Cleveland, and to New York. Early in the 1930s, I could see Boeing 247s going over, and later on the DC-3s. Sometimes when I saw the airplane going over, I would lie down in the yard, and just lay there and look at it. I thought it would really be a

---

[18] JATO is the acronym for jet-assisted take-off; despite the term "jet," the device assisting the take-off is actually a small, solid-propellant rocket.

Bob Champine next to the D-558-1. (NASA photo EC98-44613-5, originally supplied by Bob Champine).

great thing to be up there flying that airplane.

The Depression wasn't too good to me and my family. We lost our home, and I went to live with my aunt before I finished junior college. But I was valedictorian of my class in the junior college after two years. At that point, I took my physical, and was accepted in the Army Air Corps. It almost turned me down because I had malocclusion [of the upper and lower teeth]. I never could figure out how that was going to affect how I could fly an airplane, but anyway, they could see the war coming. I think they were taking everybody that was really in physical condition, and

D-558-1 in flight, still painted its original bright red color. (NASA photo EC98-44613-3, originally supplied by Bob Champine).

had the eyesight, and depth perception, and things like that to get in the program.

As Dick said, I went to New Guinea. I was in the Army Air Corps [and Army Air Forces] for five years, getting out in 1946. In the spring of 1946, *Aviation Week* and some of the other magazines I was reading were talking about the X-1, and the glide flights they were making in Florida.[19] Eventually it got to the point where the news was out that [Chalmers] "Slick" Goodlin was asking for quite a large sum of money to fly supersonically with the airplane.[20]

D-558-2 taking off with jet assisted take-off (JATO). (NASA photo E49-219).

So I wrote a letter to Bell Aircraft. I said I was an honor student in the third year of aeronautical engineering at Purdue, had 1,200 [flying] hours, 189 combat missions, and had done a lot of flying with fighters — and that I'd like to come and fly the Bell X-1. After I was flying the X-1 for the NACA, we went to the [variable-sweep] X-5 mock-up at Bell. I talked to some people who said there were quite a few individuals who had written in and said that they would like to fly the X-1. I don't know whether they were interested in the money, or whether they just wanted to fly the airplane. Scott Crossfield said he just wanted to fly the airplane. I think he wrote a letter, too.

Well anyway, I graduated from Purdue, and people came to interview us for a job. In 1948, the average engineer starting salary was about $250.00 a month. I interviewed with Ed Gough, who was Mel Gough's brother, and another engineer

---

[19] The first glide flights occurred at Pinecastle Field, Fla., before the project moved to Muroc Army Air Field (later Edwards Air Force Base).

[20] In fact, the story was perhaps somewhat exaggerated; Goodlin's contract arrangements with Bell were consistent with then-industry practices. See Louis Rotundo, *Into the Unknown: The X-1 Story* (Washington, DC and London: The Smithsonian Institution Press, 1994), pp. 126, 226-230.

who came down from the [NACA's] Lewis Lab [in Cleveland], and they accepted me. They were going to pay me $3,727 a year. That turned out to be about $140 every two weeks, which we got along with all right. It only cost 27 cents for a T-bone steak, so the salary was commensurate with what things cost.

I was in icing research in Cleveland. I don't think we want to spend a lot of time with that. I do remember we flew the B-24 once with enough ice on it that the propellers were rubbing the ice on the engine cowlings! The post or the support for the air speed indicator was underneath the airplane, and the ice was sticking out far enough on the support that even with the pitot heat on, the ice went around a little bit in front of the pitot tube. We [had to fly] the Instrument Landing System using pitch attitude rather than air speed to get back in at Cleveland.

Well, they had an opening here at Edwards, and I said I wanted to fill it. So I went to Langley and flew a lot of the airplanes that Bob talked about. The L-39 was the first airplane I ever flew in which you could push on the left rudder and the airplane would roll right, which took a little bit of getting used to.

Another airplane they had at Langley that was interesting was the [North American] P-51 [Mustang]. In a compartment on the right wing, they had set up a balance. There were airfoil models that could be put out into the wind stream. And when the P-51 was going 0.75 Mach number, the [accelerated] air over the top of the wing was going Mach 1.2. So they were getting transonic and supersonic lift, drag, and pitch characteristics of various airfoils with this model on the wing out there. This gave a little bit of feeling for what had happened to a lot of the Army Air Corps pilots that were in the P-51s and the P-38s. A fair number got into dives that they didn't pull out of. In the P-51, you could be pushing 40 pounds [stick force] at 0.7 Mach number. At Mach 0.72 it was almost neutral. By Mach 0.76 you had 160 pounds force on the stick, and you might or might not be decreasing the angle of the dive with that 160 pounds. Some pilots went on in the steeper dives and tried to trim out of it. When the air got a little denser, and the temperature went up, and the Mach number dropped off, they had [sufficient] trim in the airplane to pull the wings off. So there were a lot of unknowns that happened in the transonic speed range.

When I was in New Guinea, we had a pilot that was in our [Curtiss] P-40 [Warhawk] squadron and had an opportunity to get with a [Lockheed] P-38 photo-reconnaissance squadron that was just across the river from where we were. He ate lunch with us one day and said, "I don't think this P-38 talk is really anything serious." He said, "I'm going to go up this afternoon and really dive one." Well, later on that afternoon we saw him coming down. From the point where we first saw him 'til he hit the ground, he went into a steeper dive. I don't understand why he didn't pull the throttles back. He buried the engines about 30 feet in the ground. So it was pretty obvious that when you went into the transonic speed range, the center of lift on the wing moved aft, and that made the nose go down.

I got a P-40 up to about 32,000 feet and came straight down, and I first experienced a stick that felt like it was cast in about two feet of concrete. It just doesn't move back until you get a little denser air, and the drag increases, and the temperature goes up a little bit, and the Mach number comes back. If you throttle back, it's easy enough to pull the airplane out. This experience went on with quite a few people that flew P-51s and P-38s. Lockheed eventually put a flap underneath the front of the wing, so that if you got into that kind of trouble, you could open that flap and pull out.

But we got to Edwards here and started the Skystreak program. As Dick said, it was a beautiful airplane and really a lot of fun to fly. We were doing a lot of flying between 0.8 and 1.0 Mach number. Quite a few of the flights that I was on were very close to Mach 1. As a matter of fact, one of Dick's flight numbers shows me going to Mach 0.98-1.0. We were measuring the pitch characteristics, and, of course, the pressure distribution over the wing, and all the stability and control aspects of flying through Mach 0.80 to 0.99 — which was giving a lot of information that was pretty much needed to keep these airplanes out of trouble when they got going in that speed range.

So this was a pretty first-hand experience to indicate that there was some reason that we really needed to get into these transonic research airplanes and determine what it was about the airplanes that we were flying that would enable them to fly safely at transonic speed and into supersonic speed.

There are so many things that can happen when you start getting into the transonic speed range — especially instability of the airflow. The normal lift distribution peaks near the front of the wing. That breaks down and moves aft as local Mach 1 speeds are reached and that makes the airplane pitch down. And then there are other characteristics on some of the airplanes that might cause it to pitch up.

One of the flights that I made with the D-558-2 was a series of pull-ups at 200-240 knots. Anyway, in a pull-up, when the airplane got to a pitch-up angle of attack,[21] it would be interesting to see the position of the horizontal tail in the wing wake in a pitch-up. I expect that when the pitch attitude of the airplane was such that the downwash from the wing went over the horizontal tail, it pitched up quite sharply. Well, at 220-240 knots, it wasn't too bad. But at maybe 280 knots, when I hit that point, without my doing anything except pushing against the stick, the airplane pitched up to a stall and a snap roll. I had done a lot of snap rolls in my life. It wasn't any problem to pull out of a snap roll, but quite a surprise to be doing a pull-up, and all of a sudden the airplane's going out of control.[22]

I guess you all know that in those days, most of our data was on an oscillograph[23] that was about this wide [holds hands slightly apart]. And the distance from the baseline to the location of the parameter was an indication of your speed, or altitude, or stick force, or G force [acceleration equal to the force of gravity or a multiple thereof], or all the various things that we were measuring. Sig Sjoberg[24] told me when I was going to do this stall that was on the flight plan, "We'd like to see

---

[21] The angle of attack (AoA) is the relationship of the aircraft to the relative wind. At a 45° AoA, the aircraft is pointing 45° above the airstream.

[22] Pitch up was violent at high speeds but was much milder at moderate speeds and not noticeable at approach-to-stall speed.

[23] In the early years, an oscillograph recording system collected flight data on film for processing by female "computers" into usable engineering data. In 1967 a more sophisticated pulse code modulation system replaced the oscillograph. See Sheryll Goecke Powers, *Women in Flight Research at NASA Dryden Flight Research Center from 1946 to 1995* (Washington, DC: NASA Monographs in Aerospace History #6, 1997), esp. pp. 12-14, 45-49.

[24] An engineer at the NACA High-Speed Flight Station (later NASA Dryden Flight Research Center).

what happens with this airplane when we have the gear and the flaps down, and we're at the end point of the stall approach."

Well, this data point was a little late in the flight, and I had gotten down to about 14,000 feet. The airplane never would go real high with just the jet engine. But anyway, I got the gear and flaps down, slowly approached a stall, and pretty soon I felt like things were getting pretty loose with this machine but no pitch-up was noticeable. I thought: well, Sig wants it really slow, so we'll keep on coming back here. So I came on back to the point where the right wing dropped and the airplane started yawing to the right. I thought it was about time to stop this and recover. But it did maybe a turn of a spin. The airplane spin recovery characteristics were unsatisfactory with the gear and the flaps down, so as I was rolling into this wing dropping and yawing, I was putting the gear and flaps up. I knew it wasn't going to be long until I was going to be going quite a bit faster than I was going then. But I got the nose down, and got a little speed up. And as soon as I had the nose down and the speed up, why the airplane was flying — but I was in a very nearly vertical position.

I later checked the telemetering data, and determined that I did the stall at 14,000 and pulled out at 7,000 feet. Well, the lakebed was at 2,400 feet. I think from then on, if I was going to do any stalls, I'd be at 20,000 feet. Walt Williams was watching this from the lake. They drove the car out to be somewhere near where I stopped when I landed. And he was looking at it with his field glasses. When it slid off into the spin, he handed the glasses to Joe Vensel[25] and he said, "Here — you look!"

Another point about the D-558-1: I think the wing on the D-558-1 was about 150 square feet. And that made the stall speed a little high in some cases. I know that I was doing a clean configuration stall according to the flight plan. And I felt pretty good at 150 [knots] indicated [airspeed]. At about 149, why I had dropped 1,000 feet. And so things quit all of a sudden. As far as the high speed part of it was concerned, I flew I know at least three or four flights that went above Mach 0.97. We did several runs from a lower speed to that high nine-tenths with different stabilizer settings. And this gave us a pretty good indication of some of this tucking that I was talking about that went on with the P-38 and the P-51. However, I do remember some buffeting and some trim changes, and things like that. But I felt like it was really a pretty good airplane to fly up to near Mach 1. And I enjoyed flying the airplane. I thought it was a lot of fun.

I can't really think of any more things that are directly tied into the D-558 program. I do know that, as Dick said, in the X-1 it was a short shot. You'd get to 50,000 feet, and start down with four rockets, and maybe get up to Mach 1.2 at the most. And near the end of that time, we'd do a roll or a pull-up, or some kind of maneuver that would give them a little more information about handling qualities at those speeds. And it was not a very long time. Soon as the fuel was gone, you jettisoned the residue. Then it was a no-power flight to the lake.

They were going to have a movie called *Jet Pilot*, and X-1 number one was

---

[25] A distinguished NACA and Navy pilot, Joe Vensel transferred from the NACA's Aircraft Engine Research Laboratory in Cleveland, Ohio (later NASA's Lewis Research Center), to the not yet officially named NACA Muroc Flight Test Unit (later NASA's Dryden Flight Research Center) as Chief of Flight Operations in April 1947. He remained in the position until his retirement in December 1966.

going to be used for that. The crew came up from Los Angeles — the movie crew. They were painting the airplane, and they had some talk about people's salaries. I heard the salaries these guys were making in the movie crew, and it was more than I was making. I thought: if I'm going to be up here flying this X-1, I ought to get a little more money.

And I really don't know why. But it was just a few days after that that J.R. Clark came from Chance Vought. And he offered me a job that paid almost twice the salary, and with the bonus program, I could maybe earn four times the salary I was making. What he didn't tell me was that in the various X models of the F7U Cutlass, they crashed five airplanes and killed three pilots.[26] I didn't know that when I went to work for Chance Vought. But when I got there, from some of the flying that I did, I found out why. I worked for them for a year, and I figured I'd like to see my kids graduate from high school. So I went to work for the airlines.

I do think I probably have time to talk about one episode with this Chance Vought Cutlass that ties into the "tuck" problem that we had with the P-38 and the P-51. The F7U-1 was built with a hydraulic power control system. And if you had lost complete hydraulic power, they had a spring tab system as a backup. It was a mechanical system that could be used to recover the airplane if you lost all your hydraulic pressure. The airplane would fly quite well on the spring tab system if you weren't going fast.

So this was the bonus program I was on. I was supposed to see how fast the airplane could go and still be recovered with the hydraulic control system shut off. I made several practice runs where I shut off the hydraulic control system, but I didn't shut off the hydraulic power that opened the speed brakes. On the day that I was working for this bonus program, there was going to be a complete hydraulic failure, and I was going to have to open the speed brakes with a high-pressure air bottle. Well, the airplane would only get to maybe 37,000 or 38,000 feet. And the higher I went, the more money I was going to make. So I was trying for altitude for a long time.

At about 38,000 feet, I pointed to at least a 60-degree dive angle. And then by 29,000 feet I had slightly over Mach 1, and shut off the hydraulic system. I only had three things to remember. And I think I should have had a checklist to remember these three things. The first thing was to shut off the boost. The next one was to open the speed brakes. Well, that was easy enough. I opened the speed brakes, and nothing happened that I could tell. So instead of pulling back on the throttles, I started thinking: what's the matter with these speed brakes? So I looked in the mirror, and they were just open a little bit. And about this time I looked back in the cockpit, and I was already at 18,000 feet. And the thousand-foot needle was going around more than once a second. I went from 33,000 to 13,000 feet in less than 20 seconds. But instead of pulling the throttles back, I turned the boost back on. I was pulling about 90 pounds with one hand, and as the boost came on, I could easily have pulled the wings off the airplane. But I relaxed that pressure back to about 45 pounds as the airplane approached six Gs. I was aware that the airplane design parameters were six Gs at 520 knots equivalent air speed, and I was doing 560. But without a G-meter, I

---

[26] The F7U Cutlass was a radical twin-jet, swept-wing, tailless jet fighter. Though it did deploy aboard Navy carriers (and was the first operational missile-armed Navy jet fighter), it was not a great success and did not remain in service very long.

thought I must be pulling six Gs.

It showed on the records after I got through that I had pulled between six and six and a half Gs for eight seconds, and missed the ground by less than 2,000 feet. So if I'd have turned the boost on two seconds later, I'd have hit the ground and made probably the biggest hole that an F7U ever made. I was going 700 miles an hour at 12,000 feet. But that was a point where I thought there probably would be some pilot that would pull the wings off. There might be another pilot that would have hit the ground. And then again, there might have been a pilot that would have pulled the throttles back and avoided all that excitement. [Laughter.]

ANOTHER SPEAKER: Did you get your money?

I got part of it. But it turned out that the Navy signed off on their structures and their recovery with the boost shut off. They didn't want any more tests. The airplane had both fins bent and one rudder fluttered. And there was just a jagged piece on the post that was in the fin. I called the ground station and I said, "Well, CVA, this is Mike. I'm still here." And I was pretty glad of that. And I said, "Both fins are bent, and one rudder is gone." And Martin Collis called up. He said, "Well, I'll come up and have a look at it." I said, "Looking at it isn't going to do it any good. Just rig the chain gear,[27] and I'll come down and land it there." So that was really an uneventful landing after getting the machine out of the dive.

But I did have several thoughts. You see, thoughts run through your mind when you're in a tense situation sometimes. The first thought that went through my mind after 18,000 feet was: what's Cleo going to do with those three little kids? And the next thought I had, after I was pulling the Gs, was: I guess that engineer that designed that control arm and that hinge point there sure must have done a good job of designing the thing, because it's still hanging on the airplane.

Well anyway, I think that any of you that know anything about physiology of G forces — after three or four seconds of six Gs, most people will be at least grayed out. By the time you get to near eight seconds, most will be unconscious. And I know that I was still pulling the 40 pounds at the bottom of the dive. Because I was going back up again. The canopy completely frosted over, going from 70 below zero to 80 degrees in the Texas area there. And by the time I got back near 12,000 feet, I thought: well, I'm going fast enough and high enough, and pulled the throttles back so I could fly back to the base.

There is one other thing that maybe later on Scotty will talk about. I was really wondering why there were so many high-altitude losses of control. I know a lot of pilots — Yeager did it twice, and [Capt. Arthur "Kit"] Murray did it once. And Milburn Apt[28] — that was probably an error in judgment that they sent him that high and that fast on his first flight. But as the years have gone by, we've gotten pretty

---

[27] A runway arresting mechanism for stopping an airplane that might be damaged too seriously to stop by normal braking.

[28] Capt. Apt died on 27 Sept. 1956 after flying to Mach 3.2 in the rocket-powered X-2. The aircraft went out of control due to predicted inertial roll coupling after he became the first pilot to reach Mach 3. The rocket had burned longer than predicted, forcing the pilot into a quandary. He had either to decelerate through Mach 2.4 as planned, in order to make a safe turn but at a greater distance from the landing site than expected, or risk the predicted

well into stability augmentation, and yaw dampers, and thrusters, and things like that on airplanes. And my opinion about it is that maybe they should have done a little of this work a little lower and a little slower before they went up there and lost control. But that's probably 20/20 hindsight.

HALLION: Most of you probably heard a strong sonic boom a few minutes ago. That was a tribute, by the way, that Ed Schneider told me that he was going to make specifically for this symposium this afternoon. That was a Dryden F-18 flying through Mach 1 in honor of the D-558-2, and the accomplishments of the D-558-2 and the D-558-1 in transonic and supersonic flight testing.

Now, we're going to start this afternoon much as we did this morning. I'm going to give a quick overview on the D-558-2 program, and some of the work that was undertaken there.

I mentioned this morning that as the D-558 program went along, we had a series of two mock-up conferences. And at the second of those mock-up conferences, which took place in August of 1945, the decision was reached to split the program, so that we would have a Phase 1 that was a straight-wing aircraft and a Phase 2 that was a swept-wing airplane. How did this come about?

Basically, there had been tremendous interest in the swept wing generating in this country since the mid-1940s. In late 1944, you had had Robert T. Jones, an aeronautical research engineer at the Langley Memorial Aeronautical Laboratory, as the Langley Research Center was known in those days — who postulated the notion of the swept wing for transonic drag reduction, independently of German work. This is an important point, because I think that there's a myth that we live with in aviation history — and that is that we got the delta wing and the swept wing from Germany, and that we were ignorant of these things until we had the chance to examine the German aircraft industry. Nothing in point of fact could be further from the truth. Both the swept wing and the delta wing were indigenous American developments. And their history is a very interesting history.

In April of 1945, in fact, Jones undertook research studies on the swept-wing configuration, at the behest of Theodore von Kármán, who was an immigrant Hungarian aeronautical scientist and the scientific advisor to the Army Air Forces.[29] And they put a wind-tunnel model together — a very sharply swept model. And it confirmed that the swept wing had very good aerodynamic characteristics — up in the high supersonic range, to Mach 1.72. This is one of those classic problems I mentioned earlier about tunnel testing. You could get very good subsonic data, and

---

instability that caused his death. On this, see, e.g., Richard E. Day, *Coupling Dynamics in Aircraft: A Historical Perspective* (Edwards, Calif.: NASA SP-532, 1997), pp. 10-13, Richard Hallion, *On the Frontier: Flight Research at Dryden, 1946-1981* (Washington, DC: NASA SP-4303, 1984), pp. 76-78, and Lane E. Wallace, *Flights of Discovery: 50 Years at the NASA Dryden Flight Research Center* (Washington, DC: NASA SP-4309, 1996), pp. 54, 181.

[29] Von Kármán had been a student of the eminent fluid dynamicist Ludwig Prandtl at the University of Göttingen and later rivaled his mentor in that field of study, which included aerodynamics. He headed the Guggenheim Aeronautical Laboratory at the California Institute of Technology before becoming the scientific advisor to the AAF. See Michael H. Gorn, *The Universal Man: Theodore von Kármán's Life in Aeronautics* (Washington, DC: Smithsonian Institution Press, 1992).

you could get very good supersonic data. But in that transonic region in between, from about Mach 0.75 to about 1.25, the measurements were very suspect. Beyond that, when they were dealing with this model at about the 1.5 to 1.72 range, it exhibited very good characteristics.

Number two D-558-2 Skyrocket being launched from a Navy P2B mothership. (NASA photo E-2478).

In May of 1945, as part of the American industry's effort to study the German aircraft industry, L. Eugene Root and A.M.O. Smith, two individuals — as I mentioned this morning — who were intimately involved in the D-558 program, went to Germany as part of the Naval technical mission to Europe — NAVTECHMISSEU, as it was called. And they visited the, if you will, German Langley — the so-called Aerodynamische Versuchsanstalt [aerodynamic research facility or test station] outside Braunschweig. And they learned there of the tremendous range of work that was going on in Germany on swept-wing development. This came, if you will, as confirmation of their inclinations to pursue the swept wing. Root stayed on in Europe. Smith returned to Douglas in early August. And to show how rapidly this turned — as I mentioned, at the second mock-up conference on the D-558-2 which was held in the middle of August (August 14-17), the decision was reached to go ahead and launch a swept-wing variant of the airplane.

From the first photograph, you'll see that this was a very different beast. If you compare this with the Skystreak, as we saw in cutaway this morning, this aircraft for supersonic performance was to have a rocket engine in the back end, a so-called Reaction Motors 6000C4. That stood for 6,000 pounds of thrust from four thrust chambers. We have an example of the engine here on stage. In fact, you see the independent thrust chambers — each one of which gave you 25 percent thrust. And that would be tucked in the tail cone of the airplane. Therefore, you couldn't have a very large jet engine.

Fortunately Westinghouse, at the time, was developing a family of axial flow turbojets — the model J30 and the model that would eventually become the J34. And so the decision was reached to put a Westinghouse model 24, the predecessor of the J34, in the belly of the aircraft as well, exiting under the tail cone. So this would be a combined propulsion jet and rocket airplane. That greatly complicated, as you can well imagine, the internal fuel capacity for the aircraft. In fact, the airplane operated initially with three fuels. It operated with a liquid oxygen and water-alcohol mix for the rocket engine. It operated with jet fuel for the J34. And it operated with hydrogen peroxide to power the turbopump. So this was an airplane that was already getting pretty exotic in most respects.

Kermit Van Every was the aerodynamicist who designed the configuration of the D-588-2, working with Ed Heinemann. And if we take a look at this, it's an interesting machine. It was intended for ground takeoff and landing. There was no desire yet to air-launch this airplane. There was some thinking that maybe we'd go in that

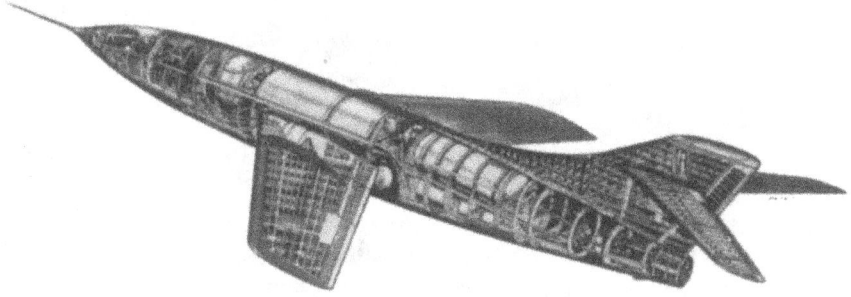

Cutaway view of the D-558-2. (Photo provided by Tony Landis).

direction, but it was far off. The airplane was designed with anhedral on the wings. In other words, they were angled downwards slightly. And they had reverse taper. They had a 10 percent thickness:chord ratio at the root, and a 12 percent thickness:chord ratio at the tip. You had Handley Page leading edge slats on the aircraft. You had wing fences, and the flaps of course.[30] And it was a 35-degree swept configuration which was relatively conservative in terms of the evolution of the swept wing at that time. It was comparable in wing sweep to the F-86 then coming along.

To ensure that the pilot had adequate control over the aircraft should it encounter transonic difficulties — to prevent the drag divergence Mach number of the wing and the tail being equal — they swept the horizontal tail surfaces at 40 degrees. And it also had a fully adjustable horizontal stabilizer, just like the X-1. The load limit on the airplane was lower than the D-558-1. Instead of the 18 G ultimate load, it had a 12 G ultimate load. It had a 7.33 G limit load, which was consistent with military fighter design practice at the time.

When the aircraft was originally designed, it had an X-1 style nose configuration. You had a smooth ogival body shape and a flush canopy. The cockpit, as with

---

[30] Slats were long, narrow auxiliary airfoils affixed to the leading edges of the wings to increase lift at high angles of attack. Fences were stationary plates or vanes projecting from the upper surfaces of the wings, substantially parallel to the airstream. They were used to prevent spanwise airflow detachment over the wing.

the D-558-1, was a confining little space. Now the first airplane flew on February 4, 1948, with Johnny [John F.] Martin at the controls. He was a Douglas test pilot of distinction, more noted for his work in attack-bomber and transport flying than high-performance aircraft flying. But a couple of deficiencies became visible pretty quickly. One of them was a very annoying Dutch Roll oscillation that resulted in Douglas eventually increasing the height and the area of the vertical fin.[31] And also, the visibility from that cockpit was pretty horrible. So the airplane was modified to have basically a Skystreak-like high-speed canopy.

It was about a Mach 0.85 airplane, straight and level on jet engine only. Very underpowered, but that would be expected, given the small Westinghouse engine.

Let's go back about ten years to 1938. If we think of 1938 and the Navy's leading fighter in 1938, it's the externally braced Grumman F3F biplane — 250 mile-an-hour maximum speed. A decade later, we have a Mach 2 aircraft flying. It's not at Mach 2. It won't be at Mach 2 for another five years. But that is how rapidly the technological change is taking place. That is the radical transformation that we're seeing in aviation technology at that time.

You know, we speak today of the fact the computational power is doubling every 18 months with computers. And that is obviously extraordinary. But if you look at this — in its own way, in a very hard-core/hard-technology sense, this is an equivalent revolution that we see taking place in terms of the profound impact it's having.

Now there were several difficulties operating the D-558-2 in its initial configuration. I mentioned that it was severely underpowered. This, of course, greatly complicated flight safety. It had some rather dangerous takeoff characteristics. Typically, it would take off with four JATO bottles strapped to the airplane to give it an additional kick in the rear on takeoff. Takeoff rolls were very, very excessive. These kinds of problems, particularly also the problem then of operating it with a rocket engine and very volatile rocket propellants at some point, caused people to begin thinking more and more about both safety and performance advantages of operating it as an air-launched airplane.

On 24 May 1949, we had the first NACA flight in the D-558-2 number 2 by Bob Champine. It was still a jet-only program. This was the aircraft then, which Bob and John flew briefly before it was returned to Douglas for modification to air launch — all-rocket air-launch configuration — in January 1950. But in this brief six-month period of flying — as John and Bob both alluded to in their presentations — it flew extensively on early swept-wing pitch-up investigations. The first pitch-up encounter was by Bob on 8 August 1949. It was a pitch-up, in a four G turn at 0.6 Mach number, to six G. John Griffith then, on 1 November 1949, encountered one that was more interesting. Severe pitch-up, a snap roll, and then a low-speed pitch-up, and a departure [from straight and level flight] in turn that was eroding rapidly into a spin.

In June 1949, the D-558-2 number three, which became NACA 145, made the first supersonic flight using both jet and rocket propulsion. Gene May, Douglas pilot, remarked, "The flight got glassy smooth — quite the smoothest flying I had ever known." I think that was an indication right there that the airplane was going to be pretty successful as a supersonic research airplane.

---

[31] Dutch Roll is a complex oscillating motion of an aircraft involving rolling, yawing, and sideslipping. It takes its name from its resemblance to the characteristic rhythm of an ice skater.

In September 1949, Hugh Dryden, who was the NACA's Director of Research,[32] recommended to the Navy that the D-558-2 be modified for air launching. Why? Three reasons — safety, performance, and research. The research attributes were that you could now compare the performance of a 10-percent swept-wing aircraft over the same speed range as the straight-wing 10-percent NACA XS-1. And you could compare the conventional airfoil cross-section of the D-558-2's swept wing with the unconventional airfoil cross-section of the Bell XS-2, which used a radical so-called bi-convex section that was then under development. That was good enough for the Navy.

On November 25, they added an amendment to the contract to modify the number two and the number three aircraft to air-launching. The number two would be an all-rocket airplane. The number three would retain its jet and rocket engine. The Navy had a small fleet of B-29s for a variety of test purposes — anti-submarine warfare research, things like this. And so a B-29, or as the Navy designated it, a P2B-1S, was set aside as the launch aircraft for the D-558-2.

We had the first air-launch of a D-558-2 on September 8, 1950 — Bill Bridgeman in the number three airplane. And then you had the first NACA flight in this particular aircraft, beginning the NACA's supersonic air-launch research program with the Skyrocket on December 22 of that same year with Scott Crossfield.

The real attention was focused less on the number three airplane, which of course was both jet- and rocket-propelled and became a maid of all work. The real attention was focused on the most glamorous of the Skyrockets, and certainly the one that has become the most famous to us, and that was, of course, the all-rocket number two airplane, which is now hanging in the Smithsonian Institution. This airplane, which received the call sign of NACA 144, had greatly increased fuel tankage over the jet-and-rocket Skyrocket. It could carry 345 gallons of liquid oxygen, and 378 gallons of water-alcohol.

If we take a look at the Douglas contract and the program on this aircraft which began in 1951, we see some interesting things and some very interesting highlights. We had the inadvertent first flight on 26 January 1951. This was a case where there was a fuel-pressure drop. Bill Bridgeman called to George Jansen, his launch pilot, and said, "Don't drop me, George." And George Jansen, his finger mashed down on the transmit button, kept intoning the countdown. Bridgeman was launched saying, "Damn it, George. I *told* you not to drop me." And the chase pilot, who was Pete Everest in an F-86, said, "You've got some keen friends, Bridgeman."[33] That's one of my favorite stories. Bridgeman recovered very adroitly, and went up to Mach 1.28 in the airplane. He noted a decrease in elevator effectiveness above Mach 1. That, I suspect, didn't come as too much of a surprise.

On May 18 — just to give you some highlights — he reached Mach 1.72 at 62,000 feet, 1,130 miles an hour, making the Skyrocket the world's fastest airplane. In June 1951, he extended this to 1.85 Mach number, 1,220 miles an hour, but experienced some very violent rolling — 80 degrees a second — causing him to

---

[32] That year he assumed the title, Director, rather than just Director of Research. See Michael H. Gorn, *Hugh L. Dryden's Career in Aviation and Space* (Washington, DC: NASA Monographs in Aerospace History #5, 1996), p. 9.

[33] Quotations in Hallion, *Supersonic Flight*, p. 164.

prematurely terminate the rocket flight with over 50 seconds of rocket fuel remaining. The problem here was, as he was going to very low pushover load factors, the airplane was becoming increasingly unstable. Bridgeman assessed this very well. And on August 7, 1951 he reached Mach 1.88 safely, using a higher .6 to .8 G pushover, as opposed to the .25 pushover load factor that he had used on his earlier flight.

Douglas then turned to the potential of the aircraft to exceed the world's altitude record, which was held by the balloon *Explorer II*, going back to 1935 — a 72,395 foot record. Bridgeman on 15 August 1951, reached 79,494 feet, making the Skyrocket both the world's fastest and highest airplane. I think this is a tremendous tribute to Bridgeman as a pilot, and to Ed Heinemann as the designer of the aircraft. The airplane, in fact, when you took a look at it, had some significantly better performance than its designers had predicted. In fact, its supersonic drag was actually less than what people predicted at the time.

D-558-2 number two returning from a research flight with an F-86 flying behind it as a chase aircraft. (NASA photo E-3996).

If we take a look at a couple of classic photos from this period, Bridgeman developed a very close association and friendship with Chuck Yeager, who flew a lot of the F-86 chase missions. And this is a very evocative photograph, I think, of the D-558-2, drifting down from a research flight with Yeager in the F-86, speed brakes deployed, coming down behind him.

Now, for the NACA's part: you know, if 1951 was the time in which Douglas was exploring the high-speed realm with the all-rocket number two airplane, the NACA's part — working on the D-558-2 number three — began basically what would become a two-year program here. And Scotty will certainly be talking about this, and Stan as well, involving basic aircraft handling qualities and evaluation of various flap, fence, and leading edge devices on the aircraft.

In 1952 and 1953, the NACA shifted to examining the high supersonic behavior of the D-558-2 number two. We have here, I think, another evocative photograph. This is 144 in its prime on the lakebed. And you can see how futuristic it really looks. And these were really Scotty's glory days in the Skyrocket. Some highlights here: August 5, 1953, he reached Mach 1.878; August 21, the Navy borrowed the airplane for some high altitude and high speed flights. It was hoping — frankly — to break Mach 2. It didn't happen. Marine test pilot (Lt. Col.) Marion Carl nevertheless

D-558-2 number two on the lakebed. (NASA photo E-1441).

distinguished himself. On 21 August, 1953 he reached 83,235 feet, an unofficial world's altitude record. On 14 October, 1953, Scotty reached Mach 1.96. The airplane had boosted performance at this time, due to a rocket nozzle extension on it.

And at this point, the High-Speed Flight Research Station now requested and got Hugh Dryden's permission to attempt Mach 2. Herman Ankenbruck devised the flight plan. Scotty would basically climb to 72,000 feet, do a pushover, and reach Mach 2 in a shallow dive. The plane was extraordinarily prepped for this. Scotty will go into that in much more detail than I will. And on 20 November 1953, as I think we're all aware, he reached over Mach 2 — 1,291 miles an hour at 62,000 feet — the first Mach 2 flight, which was a tremendous accomplishment — both reflecting on Scotty's abilities as an airplane driver and the design of the airplane. This was undoubtedly the high point of the D-558 program.

We have to recognize that a lot of people made this thing come together. I'd like to talk about some of these. These are the P2B as well as D-558 crewmen. And, of course, supporting these people were folks here at the Center on the ground — the maintenance staff you know. The test pilot in this process is merely a singularity, so to speak — the tip of the spear. But that spear is forged and wielded by a great number of other people.

Now if we take a look at the twilight years in the Skyrocket program from 1954 through 1956, the last flight taking place on December 20, 1956, by Jack McKay — and we have Jack's son John with us today. If we take a look at it, these were not years in which things went necessarily very smoothly, although they were undoubtedly extremely productive.

Stan Butchart, Neil Armstrong, and Jack McKay had a very up-close and personal encounter with a near disaster in 1956 that I think Stan will be giving us a great deal more information on — when they had the number four engine run away with them, shed its prop, and do some serious damage to the launch aircraft, and indeed pass right through the space where a few seconds before Jack McKay had been in the D-558-2 before it was jettisoned in an emergency. That was about the most dangerous moment, I think, in the entire D-558-2 test program.

We had then the fruition of work on the D-558-2 number three's pitch-up investigations, which resulted in some experimental design changes to the airplane, some of which were quite promising, but which didn't pay off. We had, for example,

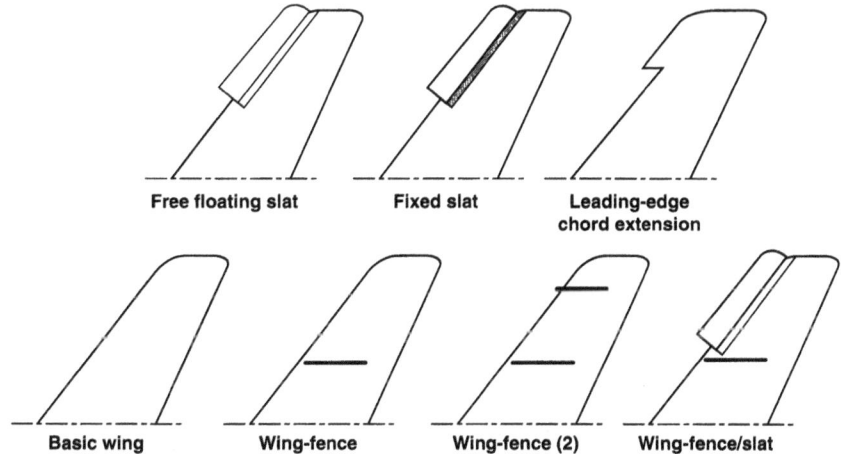

D-558 wing configurations.

the effort to explore behavior with leading edge slats open. And indeed, fully open slats did work to a great degree. They eliminated pitch-up, except between 0.8 and 0.85 Mach number. However, surprisingly, a sawtooth leading-edge extension from which much was expected actually, in Scotty's views, aggravated the pitch-up problem significantly. And so it proved of no value whatsoever.

A little-known aspect of the D-558 program is that after going through this pitch-up program, it embarked on a number of investigations of external stores, looking at the drag of external stores on aircraft at transonic and supersonic speeds.[34] Now, this is extremely significant work. Because if we think about the Mark (Mk.) 80 family of stores — the Mk. 82, the Mk. 84, other bomb shapes, drop tank shapes — that we live with today, that basically is an outgrowth of the D-558 program. The D-558 took these shapes, which were experimentally developed by Douglas, and refined them to the point now that we could operate strike aircraft at long distances with streamlined stores with significantly less drag than the kind of clunky bomb shapes and tank shapes we were operating with that were basically holdovers from the World War II and immediately post-World War II era. The Mk. 80 store shape, which was applied generically then to a whole family of shapes for both tanks and bombs, was really quite a remarkable accomplishment. And the D-558 played a role in it, both here for bomb shapes, and for tank shapes as well.

At this point, I'd like to terminate my little presentation on the raw history, if

---

[34] "Stores" were such things as external fuel tanks or bombs.

you will, of the D-558-2. And we will move from this point on to discussion of actually operating the D-558-2 aircraft. So at this point, I'm going to introduce our two very distinguished personalities that we have here this afternoon to talk about these: Scott Crossfield and Stan Butchart. So, Stan — first we'll start with you.

Stan, of course, was out here for a number of years. He retired from Dryden in 1976, after a 25-year career in research aviation. Born in New Orleans, 1922. Served as a naval aviator in World War II. Graduated from the University of Washington with a Bachelor of Science Degree in Mechanical and Aeronautical Engineering in 1950. In fact, Stan and Scotty were in the same Guggenheim Aeronautical School at the University of Washington. And Stan began his career with the NACA in 1951, a year after Scotty.

His experimental flight career included piloting the X-4, the X-5, both the Skystreak and the Skyrocket. He flew the B-29 that launched the X-1A, and then the B-29 — the P2B-1S — that launched the D-558-2. They're obviously not the same B-29. And also then flew the KC-135 tanker out here, and the F-100A.

D-558-2 number three with a bomb shape under its wing. (NASA photo E-1161).

But there're a couple of other things about Stan that I think we need to mention. I'm sure a lot of you are aware that one of his best buddies in all the world from Torpedo-Bomber Air Group VT-51 on the *San Jacinto* back in World War II was a fellow by the name of George Bush, who went on to bigger and better things. Had Stan profited from that wise counsel and followed a different career path, think of how different the world might be today!

Stan has another distinction that I just learned about last night. And I got to thinking about this. It's really quite interesting. Stan flew the Grumman Avenger, which was a big, hefty torpedo bomber. It was called the TBF when it was built by

Grumman. Grumman couldn't meet the Navy's requirements for mass production of the aircraft. It concentrated instead on fighters. And so most of the Avengers were built by General Motors and were called TBMs. It was a maid of all work for the Navy — did some tremendous work — anti-submarine patrol, attacks on islands, did a lot of basically 500-pound bombing, things like that. In other words, it served primarily in roles other than what it was originally intended for, flying as a torpedo bomber.

Well, Stan is one of the very few people who not only learned to drop torpedoes, as he was becoming a naval aviator and proficient in operating the Avenger. But he actually dropped a torp in combat. In fact, he dropped four of them, I believe. And one of these was against the Japanese carrier *Zuikaku* (which was one of the six that struck Pearl Harbor in 1941) during the Battle of the Philippine Sea. And he's too modest to state with certainty that he got a hit on it. But it absorbed several torpedo hits in the Battle of the Philippine Sea. And I'm certainly willing, for the record, to accord him credit for it. So, Stan, you played a role in avenging Pearl Harbor. And I think we all owe you a tip of the hat for that.

Now I'd like to introduce also a very good friend, Scott Crossfield — a legendary figure in aviation certainly, and an individual that I have a fond affection for. And I'll explain why a little bit later.

Scotty joined the NACA in June 1950. If we take a look at the roster of airplanes he flew, it's sort of a who's who and a what's what of research airplanes — the X-1, the X-4, the X-5, the XF-92A, the D-558-1 and 2. He had 87 rocket flights in the X-1 and the D-558-2 aircraft, plus 12 flights in the D-558-2 on jet power only. He flew a number of modified service aircraft. He did zero-G studies in the F-84, roll coupling studies in the F-100 and the F-86. I think he even cracked a vertebrae at one point, if memory serves me right, in the F-100 in some of the roll coupling work.

He made aeronautical history obviously on November 20, 1953, with his Mach 2 flight. But then he left the NACA in 1955 in an act that was pretty selfless. He was very concerned about the future of the X-15, which he could see was a potentially milestone airplane. And he was very concerned about some of the glamorous hangar queens that had come along, that had actually had some serious difficulties — the Bell X-2 and the Douglas X-3 being notable examples.

And so he went to work for North American Aviation to shepherd the X-15 through its development and through its contract-to-flight test program. And I think the fact that he did that explains in large measure why the X-15 was the tremendous milestone airplane that it was. During his flight testing of the X-15 with North American, he flew the airplane 14 times, made 16 captive flights additionally in it, reached a max Mach of 2.97, 1,960 miles an hour, at a max altitude of 88,000 feet.

Then he did something that I really find interesting. In 1960, he published an autobiography called *Always Another Dawn*.[35] And I realize he wrote this when he was age 39. And what's very interesting is if you take a look at Scotty from that point on, he ought to really start thinking about working on volume two. Because from that point on, he continued to do a tremendous amount of work with North American on various programs — the Hound Dog missile program, Paraglider, the Apollo Command and Service Module, the Saturn booster.

He went to work as an executive with Eastern Airlines. He went to work with

---

[35] Subtitled *The Story of a Rocket Test Pilot* (Cleveland, OH.: World Publishing Co., 1960).

Hawker Siddley Aviation. Served as a technical consultant from 1977 to 1993 on the House Committee on Science and Technology. And Scotty, I think there're many — myself included — who might argue that given what you did with that Committee in that period, it actually may be among the very most significant things that you did. Because you helped keep the Congress straight on aeronautical issues for quite a while there.

We're talking to a man here who's a Collier Trophy winner for 1961 — the Clifford Harmon Trophy for 1960. But the thing that I really remember Scotty for is the fact that when I was an undergraduate at the University of Maryland, I did a senior thesis on transonic and supersonic research airplanes. And I'm kind of embarrassed by it now, frankly, when I flip back through it. But then, with the height of ego, I sent it off to Scotty and said, "Give me your thoughts on this." And he did! And not only that, but they were polite, which amazes me even more, given what he was reading.

And then he took the time to invite me down to meet with him. So I met with him in — I think it was the summer of 1970. And he spent the better part of a day going over this thing, page by page. So Scotty, I personally tip my hat to you, because I owe you a lot for that. And I appreciate it very much. I'm delighted that I'm able to have a chance here today to introduce you to this symposium.

So Stan, we'll start with you. You're up. And the subject is B-29 or P2B-1S launch operations in support of the Skyrocket program. And then we'll follow with Scotty talking about the events leading up to his Mach 2 flight. [Audience applause]

BUTCHART: As Dick said, I had flown both the Skystreak and Skyrocket. I was mainly asked to talk on the B-29/P2B-1S mothership operations. But since I did fly the other two, I'd like to make a couple of comments there.

The Phase 1 was the first research plane that I flew. And I considered it a fun airplane to fly. It was small. And it was just fun to fly. But there's kind of a little story that goes with the day I came down here from Boeing — flew down to be interviewed by Joe Vensel for a job. And we spent the day watching an X-1 ground run, and looking at all the airplanes in the hangar, and crawling in them. And when he got through that afternoon, he wanted to take me up front and introduce me to Walt Williams — the big boss. And the only thing Walt said when he met me, "Will you fit in the Phase 1?" And I said, "Yes, sir." "Okay. You're on."

But the Phase 1 was interesting. As Dick mentioned this morning, the air split and went down the sides. The cockpit was only 22 inches wide, straight down the sides. You flew it with your elbows in, and the wheel between your knees, and crunched down. Your helmet was up into a tight canopy. We had a chamois skin on our helmets to keep from scratching the inside of the plexiglass. There was a double layer — glass and then plexiglass with air in between to keep the frost off. And if you turned your head a little bit to try to see out to a chase or wing tip, your head would get stuck, and you'd have to suck it back down to get forward again. If you ever had claustrophobia, that was the airplane to get it in.

But a couple of other little interesting things that happened in it. I mentioned one the other day to [someone] — I think Bob was the other culprit along with me. But most of the flights made in that airplane were with wing tanks — tip tanks. You took off. And when you got to 40,000 feet, your tip tanks were empty. And you could jettison out over PB-6, one of the bombing range targets out back here. By the time I got my turn to fly the airplane, Joe Walker had used up all the tip tanks. So all of my flights were made just with internal wing fuel. And it was only 202 gallons. You had a little odometer on the instrument panel. And they'd set it at 202. And as you started the engine, it started clicking down — two gallons at a time. By the time I'd get to 40,000 feet, I'd have enough fuel to do one or two runs, and it was time to head home.

Well, one day I came back. And I guess I stretched it a little bit. And I landed. Joe Vensel walked up and looked in the cockpit, and there was [a reading of only] eight gallons showing. And all he said was, "There was another pilot out here ahead of you. He came back with 12 gallons, and I grounded him." And I think that was you, Bob. But it was going fine. Eight gallons was great.

The other thing that was an interesting story on that little airplane — you saw pictures of it earlier. I'd forgotten that the canopy opened from the back forward. They'd take it out to the south lakebed where we made all our flights from. They'd get set up. And you'd crawl into the thing. And they'd help you strap in.

And Raczkowski — Tom Raczkowski — was the crew chief. He'd get on one side on the stepladder. And Andy Hyland was the inspector. He was on the other side. And as you made your engines start, they were watching to make certain you didn't over-temp it, or didn't do anything wrong. Anyway, when they'd get through, why they'd close the canopy. You had two handles to lock it with. I locked it, and I thought I was all set to go. Then they finally motioned me to open the canopy. They didn't like something. I guess it didn't fair in with the fuselage the way they had expected it to.

So they opened the canopy. And there was an air tube that blew hot air out into a delivery tube in between these windowpanes to keep the frost off. That hot air was blowing in my face. So while they were working on it, I had my hand up over this tube to keep the hot air off. And all of a sudden, Raczkowski decided to close that canopy. My thumb was still there. And oh man, you know, you jump and wail. They saw something happened, and they said, "You okay? You want to go?" It didn't hurt, so I said, "Okay." By the time I got to 40,000 feet, that thing was going thunk, thunk, thunk. And boy I wished I was on the ground.

But one other little item on both the Phase 1 and the Phase 2: all the airplanes nowadays — everybody is proud when they get an airplane that it has "zero-zero" escape capabilities. In other words, you can be sitting at zero air speed on the end of the runway [at zero altitude], and punch the eject button, and go out and make it. Well, as I think back, we had zero-zero on the Phase 1 and Phase 2 — in reverse really. There was a flight envelope. There was a little spot here that it was safe to get out in. And it wasn't an ejection seat. You got out by pulling a handle. And the whole nose fell off. You pulled another handle. That released the back rest. Then you crawled out the back. And this envelope was so small in altitude and speed that we would look at that information and put it in file 13 and go ahead and fly. But that was our "zero-zero" in reverse.

Well, I'd better get on to the main thing I was going to talk about, and that's our launch operations. When I came to work here, as Dick mentioned, I'd always been single-engine.[36] And I'd been here a week or two. Scott was going to check me out in the twin-engine C-45 that we had. We went out to the end of the runway. A kid

---

[36] That is, had flown single-engine airplanes.

named Don Turndrup (I think his name was) was flight engineer. He was sitting down between us. Typical fighter pilot, the first thing Scott does is jam both those throttles forward, and we do a 45 degree turn. He got it straightened out, and away we went. He finally got me checked out I guess, and I flew it for two or three weeks. Well, there's a moral there. Never have a fighter pilot check you out in a multi-engine airplane. They can't.

I had forgotten the date of Bridgeman's last flight. You say the 15th [of August, 1951]. Well, two days later on the 17th, George Jansen, the B-29 pilot for Douglas,[37] called Joe and me and said, "Come on down. We're going to take a flight in the B-29." And we went out and flew for an hour, made a couple of landings on the south lakebed. We were B-29 pilots. You know, nowadays you're months, or weeks anyway, going to ground school.

As I looked back in my log book, I noticed that that was on the 17th of August. A couple of days later, Joe and I took it out for a fam[iliarization] flight on the 21st. On the 22nd of August, we made our first drop flight with Walt Jones in the 145 — the one with the jet engine.[38] So we got underway in a hurry. And I think we made three or four flights with Joe in the left seat. Then he turned it over to me, and I had it for the next six years.

I made the following chart up to show the extent of the flying we did from '51 through about '56 with the rocket airplanes. I've included the three — X-1A, X-1B and X-1E — just to show the number of flights we made.

Drops 1951 – 1956

Pilot	D-558-2			X-1			Total
	143	144	145	X-1A	X-1B	X-1E	
Scott Crossfield		43	19				62
Walt Jones			5				5
Joe Walker		3	1	1		19	24
Jack McKay	1	11	9		11	3	35
Neil Armstrong					2		2
Pete Everest		1	1				2
Marion Carl		5	2				7
Al Boyd			1				1
Totals	1	63	38	1	13	22	138

Aborted Drops 1951 – 1956

Pilot	143	144	145	X-1A	X-1B	X-1E	Total
Scott Crossfield		7					7
Walt Jones			1				1
Joe Walker				1		8	9
Jack McKay	1	1	1		7	1	11
Neil Armstrong							0
Pete Everest							0
Marion Carl			1				1
Totals	1	9	2	1	7	9	29

Grand Total 167

---

[37] George Jansen was a noted Douglas test pilot. He had been a B-24 pilot in World War II and was a veteran of the Ploeşti raid of August 1943.

[38] Douglas D-558-2 #3 (bureau #37975, NACA 145).

Scott has the largest number of flight drops. Jack McKay was next in line. These were just the flights that I made. Joe Walker made a few while I was on vacation. And the bottom half is the aborts that we made. And by aborts I mean instances when we had to bring the rocket airplane back down with us. If I dropped it and the engine didn't fire, that was their fault. At least I got off the hook.

But it was a busy, busy time. We were flying through the summer of '56 — '55 and '56. And we were operating almost six rocket airplanes at the same time. I was making two, sometimes three, flights a week on either a Skyrocket or one of the X-airplanes.

I was going to show our typical daily operation. And Dick mentioned something about what he thought the scariest part of the operation was — that accident we had. And I think *this* was the scariest part. [Shows photo E-1013.] You'd put the B-29 on

D-558-2 number two being positioned under its Navy P2B mothership being elevated on hydraulic jacks. When the Skyrocket was in position, the P2B would be lowered so the D-558-2 could be attached to the bomb bay of the "mothership" for climb to altitude before being launched. (NASA photo E-1013).

jacks. And you'd get it so high in the air you just weren't certain whether it was going to make it or not. We were controlled by wind. We couldn't do this in more than about five or six knots of wind, I think. Anyway, it was pretty low. And once we moved up to this facility[39] from South Base, we found we could get in a hangar, and get it between the beams, and jack it up high enough to get the Skyrocket under it. And you didn't have to worry about the wind then.

After they got the thing loaded, they would tow it out to the area where we had the storage tanks for the liquid oxygen and the water-alcohol and peroxide. And it

---

[39] To the present location of the Dryden Flight Research Center from the old location on South Base.

Crew next to P2B and D-558-2. From viewer's left to right, standing, Donald Hall, Dick Hanna, Bill DeGraff, Joseph L. Tipton, Charles Russell; squatting, Joe Walker, Stan Butchart, Dick Payne, Walter P. Jones. (NASA photo E-677).

was out just on the south side of where the big hangar is here now — to load the propellants on board.

There was one interesting thing that happened about '55, I guess. When we moved up here in '54, there was no taxiway between the NACA and the Air Force. It was a year later before they built that. So we would load, and then take off on the lakebed. We used the lakebed a lot anyway. But I think it was in the summer of '55 then. They were going to build the hydraulic hoist that you have out there yet. First they had to dig the concrete out — pretty big area. And then the ground was so hard that they would dig in, and put some dynamite in, cover it with plywood, blow it, and get in and dig some more. And nearly every day we'd hear a dynamite blast go off.

And one day in the summer I was getting ready to go on leave. And Vensel says, "Well, can you stick around until we get this flight off?" "Sure." So it was on a Friday, I think. We were standing by his office, and the B-29 was loaded. It was sitting out in front of us there. And all of a sudden, kaboom — a bigger blast than normal. I looked up and a piece of plywood was flying through the air. It went right through the elevator of the B-29. So I said, "Joe, I'll see you in a couple of weeks." And away we went. But once we got those hoists put in, why, I think you're still using them to lift the vehicles up underneath the B-52.

Well, we got loaded. Now we're back to that crew again. I wanted to show a picture of the crew. And I don't know if I can even remember all the fellows that were there. But the fellow on this left end was one of the fellows that served as flight engineer for me. And I think on the far side I see Dick Payne, who was crew chief on some of the X-airplanes for us. I'll take a peek. Yes. I think that's Joe Walker and Walt Jones both there with us. If you noticed on the front of the airplane, of the B-29, there were lots of patches of Skyrockets.[40]

---

[40] Each patch indicated a separate drop.

Anyway, we got the thing loaded and ready to go. The typical operation, the way it went was that if Scott was wearing a pressure suit on a pretty high altitude flight, he'd come aboard the B-29 with it partially on, not fully zipped up, and not in it completely. I'd get the airplane — the B-29 — airborne, and oh, maybe we'd climb out 7-, 8-, 9,000 feet. He could finish zipping up into his pressure suit, and it was time for him to get back in the rocket airplane. Two crewmen would go back with him.

Once they got the canopy closed, and the radio hooked up, and were happy with it, they had a switch they could throw. It would light a light on my instrument panel that would say "ready for drop." And this told me that from then on, any time we had a problem in the '29, our gentlemen's agreement between us was, I'd get rid of him. If he had a problem [in the drop airplane], I'd get rid of him. We'd fight our battles by ourselves. But fortunately we didn't do that on one occasion and, I think, saved a pilot. And another time we did it and saved another pilot.

But those little '29s took a beating. You were at climb power for at least an hour. And if Scott wanted to get a little higher — 34-35,000 feet, we spent another thirty minutes on the last 3-4,000 feet. And the airplane was at climb power all this time. And you know what that does to an engine. Well, there were a few times when the engine would fail by the time I got to altitude — swallow a valve, or something would go wrong.

But the day I want to talk about was March 22nd of '56. Neil [Armstrong] and I were flying the '29. Jack McKay was in the rocket airplane. And just as we got to altitude — around 31,000 feet, somewhere over Palmdale, the number four engine quit. It just quit running, firing. I turned around. Well, I used to kind of let the co-pilot do a lot of the flying. And I would direct him where we wanted to go. Then I could turn around in my seat and kind of watch the flight engineer and the rest of the crew. And I asked [Joseph L.] Tipton, "Well, did you try cross feeding?" "Yep." "Did you try this?" "Yep." We went through two or three things. And nothing was working.

So I thought: well, no sweat. We've done this before. We'll feather it.[41] And I hit the feathering button. And it looked like it was stopped. The blades looked like they were stopped. And all of a sudden the engine started winding up again. And this particular '29 had a separate tank for feathering. So you could feather once, unfeather, and feather again. And I knew I had two left. And I think about that time Jack called me. He said, "Hey, you can't drop me." A valve down at his side that he jacked up some of the nitrogen pressures for the engine with, broke. He said, "I felt it break in my hand." I said, "Okay." And I think we hit our six-minute point by then. I had picked a six-minute point opposite of where we wanted to drop. Two minutes out, two minutes in the turn, two minutes back.

And we started through that procedure. I hit the feathering button the second time. Same thing happened. And I thought: well, we'll make it on around, and get this drop over with. And part way around, I guess that's when Jack called me and said, "Don't drop me." Well, about that time I had hit the feathering button the third time and ran out of fluid again — the last time. And I called Jack and said, "Jack, I've got to drop you." I told Neil to push over. We had to get in a dive — to get up to

---

[41] To feather a propeller is to rotate it so that the blade is parallel to the direction of the airflow so as to reduce wind resistance.

about 210-220, so they wouldn't come out in a stall. As soon as the needle got it pointed around headed back towards the lake, we were up to speed. I reached up to pull the emergency handle. I had watched them test that every time they ever loaded. It was a T handle on the dash. I reached up and pulled, and nothing [happened]. The other way was to hit two toggle switches, and then pickle it off. And that worked.

Series of photos showing damage to the P2B March 22, 1956, when the number four engine exploded and shed its propeller. (NASA photos E-2200, E-2203, E-2210, E-2213, E-2221).

And away he went. And just a few seconds after he departed, that engine blew — big time.

I could remember seeing the page in the handbook that said if you were above 20,000 feet, and you couldn't slow it down below 120, you were going to have a centrifugal explosion. Well, we did! And Neil said it looked like the kitchen sink going by. It was the nose dome off the B-29. And the blades went in all four directions. Unfortunately, one of them went right through us. It cleaned off that engine completely, as you can see. One blade went through the bottom of the number three engine there. And I don't know if you can see the slot in the fuselage. It went through the bomb bay, right where Jack was sitting. And it hit the number two engine on the other side.

Well anyway, when I heard that boom, I thought: well, we're home free. We got it made. About that time, I reached up to help Neil fly the airplane, and my wheel was loose. I just [turned it and] nothing [happened]. I looked over at Neil and said, "You got lateral control?" "Yeah, a little bit." And he had that much free play in his wheel. [Shows a small space with his hands.] It had cut part of his aileron cables. And the frayed cables were sticking out. And he was dragging those through a fairlead. And they'd get caught once in awhile.

Well anyway, while we were wrestling with that, Tipton turned and said, "Butch, you've got to feather number three." And I said, "Why?" Well, the shrapnel had hit the side of the airplane in so many places that one had cut right through our fuel line, our throttle cable, oil pressure, everything on the engine. So we essentially lost control of [engine number] three. And it feathered all right. Now you can't see there. But it actually hit the bottom part of the number two engine. So you know that blade was traveling going through there. And fortunately number three feathered all right. But that left us with two engines on one side. The only nice thing was we were at 30,000 feet. So we glided out around Boron and came straight into the lakebed.

Then Neil kind of got in an argument. He said, "You'd better get your gear down." I said, "Wait a minute." "We're getting closer. Better get your gear down." Well, nobody had ever made a 30,000-foot approach to the north lakebed in a B-29. He kept thinking we were going to overrun it. And I wanted to make certain we got there. Because I could only use [engine] number two. Number one had too much torque [being further out on the wing]. And both of us on the rudder could not hold it. Anyway, we landed with both of us on the elevators, and both of us on the rudder, and he on the ailerons.

I think we made our last flight on the 20th of December of '56. Jack McKay made it in 144. After that the airplane sat for quite awhile. This was just some time before the Navy came to pick up the various airplanes to take them to the museum. In fact, that was the first time I ever got to see the nose detach to see how that ejection system worked. On the Phase 1, there were two bomb shackles built vertically into the bulkhead. When you pulled a handle, you released the bomb shackles. The Phase 2 was a little different. It had a wheel quite similar to the bank vault, where it pulled pins in from the side. And after everybody had flown the airplane, we finally got to see how it worked.

But they sat like this for awhile. And then Neil and I took the P2B-1S over to Litchfield Park in Arizona, which happens to be the Navy's storage field for airplanes quite similar to what Davis Monthan is for the Air Force. And there's one more little part of that story.

We took it over there on the 5th of August '59, figuring that's the last we'd see of it. It would be melted into pots and pans, like everything else. And about 20 years later, Neil sent me a little newspaper clipping with a story about some guy with a lot of money who had gone in there. And the airplane was still sitting there. He purchased it, and brought in a crew to refurbish it and go over it. And it had zero time engines on it when he took it over. But they still had to go through them all. Anyway, they spent a lot of money going through it, fixing it up, and getting it ready to fly.

He hired some retired colonel who had flown B-29s to fly it for him. And they were going to make a couple of local flights around Litchfield before heading for Oakland. And when he came back from the first flight, he asked him how it was. And the colonel said, "Fine. Except I had to hold that wheel over most of the time." So they went through it, and checked the trim, and checked the rigging, the whole nine yards. And they went out and flew it again. The same thing happened. And that time they really got into it — took all the inspection plates off, and just did a real thorough inspection. And what they finally found was that all the years it sat there, some of the inspection plates were off the bottom of the wing, and birds had been living up in there. And there was about 800 pounds of bird dung out in that one wing. And that's what he was holding up. And they cleaned that out. And the airplane flew over to Oakland.

I never did get to see it. But I had heard it was on the West Coast going to air shows and what not. Then I lost track of it, and a few years ago, at one of our squadron reunions — I think in New Orleans or Pensacola — this fellow that came to our reunion every year and owns a TBM said, "Hey, I found your B-29." I said, "Where is it?" "It's in Florida." Some guy by the name of — is it [Kermit] Weeks? He buys a lot of airplanes. He's got a lot of money. He purchased the thing. The fellow that was telling me said they're going to use it for a static display. Then this last summer when we were back there, he said no. It got damaged in that hurricane a couple of years ago. But he is going to rebuild it as a flying machine. So I'm looking forward to seeing it again. But he went back. Took out the cutouts that we had in it. Put bomb bay doors on it. And put it back so it looks like a B-29, P2B-1S.

Neil used to get the biggest kick out of taking people back — to the back of the airplane and showing them where it said: NAVY P2B-1S. It had been painted on there years ago. Even when you take the paint off, it's still kind of etched into the fuselage. And he thought that was neat. So that was my experience mostly with all the years of making those drops. And Scotty, I think it's probably all yours now. [Audience applause]

SCOTT CROSSFIELD: How sweet it is to be last. You were always late, Butch. But that's all right.

This is kind of a nostalgia trip for me. But I'd like to make one aside while I get up here. I'd like to dedicate my part of this 50th anniversary celebration for the Phase 2 to Walter C. Williams. [Audience applause] Walt Williams probably had more to do with advancing aeronautical and aerospace arts in the 20th century than any other ten men, as far as I'm concerned. He started out with a crew of 12 people here at Muroc. And he was with that program — all the programs that NACA, and NASA, and many that industry did, all the way up through the Space Shuttle. He made the operational go/no-go decisions for every one of those, and had quite a part in keeping some of them from becoming national disgraces. Frankly, I'll say that while I have a high regard and respect for Wernher von Braun,[42] Walt Williams has been an order of magnitude bigger contributor to what we've done in space than von Braun. So Walt, if I do well, this is to you. If I don't, well. . . . And I'm not sure he's up there, frankly.

You know, there is no history, only biography. If you stop and think: if we ever talk about anything being done, it's done by people like these people down here, who have proved that anybody who can read without moving his lips can fly an airplane.

Another key figure was Jack Russell. Jack Russell probably did more rocket flights than any other 20 men in the world. He was with Bell on the original X-1. He came and worked for the Air Force. And then he came to NACA, stayed all through NASA. And he was with all of the rocket [airplane] flights, I believe, that were ever made. And he was one devil of a good rocket mechanic, technician, and all-around guy. So this is to Jack Russell.[43]

And then there's another guy that I'd like to pay a little tribute to. And that is the man who brought the United States into prominence with rocket engines. And that was Captain Bob Truax of the United States Navy. He started in 1937 working with the Navy building rocket engines. And actually this family of engines built by RMI [Reaction Motors, Inc.] up in New Jersey was a Navy part number in 1944, before any Paperclip,[44] before any Germans came over here and claimed Goddard's invention,[45] and everything else. Those engines were in the inventory for over 35 years. To my knowledge, we never lost an airplane due to the failure of or a problem with the engine, per se. And I'd like to give Bob Truax a little boost on this sort of thing.

JOHN GRIFFITH: I think Gerry Truszynski[46] should be mentioned in development of the instrumentation and the capability of bringing back the data that could be

---

[42] Wernher von Braun, of course, was director of NASA's Marshall Space Flight Center from its inception until 1970 and in that capacity, headed the team that developed the Saturn family of rockets that carried 12 astronauts to the Moon.

[43] John W. Russell worked on the XS-1 for Bell and then became crew chief for the Air Force on Chuck Yeager's XS-1, "Glamorous Glennis." He came to work for the NACA High-Speed Flight Research Station in 1950 and for many years headed the rocket propellant group at what became the NASA Dryden Flight Research Center, retiring March 11, 1977.

[44] Operation Paperclip brought many German scientists and engineers to this country after World War II.

[45] Robert H. Goddard was an American rocket enthusiast who worked with only a small crew of technicians. He managed to invent many of the technologies used on later rockets, but because of his secretiveness, almost all of them appear to have been reinvented by others. Thus, although he is considered by many to have been the father of American rocketry, it is arguable that his actual influence was slight.

[46] Gerald M. Truszynski was Chief of the Instrumentation Division at the High-Speed Flight Station and its predecessor organizations. He worked on the XS-1, D-558 and other early research aircraft and was responsible for setting up the High Range used for the X-15 flights before he moved to NASA Headquarters to set up the Project Mercury worldwide tracking network. See oral history interview of him and Hubert M. Drake, Nov. 15, 1996, in the NASA Dryden historical reference collection.

analyzed to produce the reports that were the product, and the result, and the purpose for what we were doing.

SCOTT CROSSFIELD: The D-558-2, as I knew it — I never flew it as a ground takeoff airplane — was the airplane that wrote the book. The X-1 air-launch techniques had proven to be good as a poor man's first stage to get rid of a lot of the energy requirements at the front end of getting an airplane to altitude. That's why they went to air launch for the D-558-2, and for many of the other reasons that my good friend Dick Hallion discussed.

The air that we fly in doesn't like high sweep angles. It doesn't like severe taper ratios. And it doesn't like low aspect ratios. And the D-558-2 had a little bit or a lot of every one of those. And it was classic in what it did as a swept wing. And that's primarily the part I will discuss, as far as the handling qualities are concerned. The tips of [the D-558-2's] wings tended to stall before the roots of the wings.[47] And if

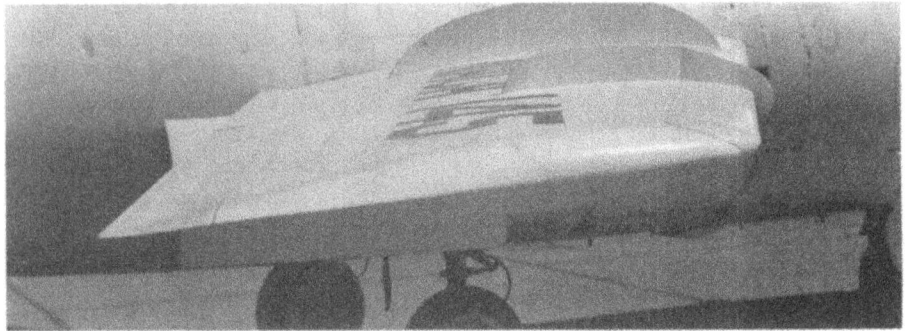

Wing fences on a D-558-2. (NASA photo E-580).

Wing slat on a D-558-2. (NASA photo E-816).

that's aft of the center of gravity (CG), the airplane wants to pitch up. And the pitch-up of the swept wings was the only characteristic that we didn't like. Everything else was in our favor. It was low drag, had excellent supersonic characteristics, and many other things.

So the D-558-2, probably with the group of the pilots here, must have done thousands of pitch-ups, with almost every device known to man on the wing of the airplane. And I'm going to discuss a few of those right now. What we thought was a massive bureaucratic operation in our day was to get one research airplane in the air,

---

[47] Stalling consists of flying at an angle and speed such that the wing (or parts thereof) experiences a separation of airflow and loses lift.

and to support the pilot down there in front with all of those great people of NACA on the South Base. And incidentally, the whole organization at that time was only 70 people, including the janitors. Can you imagine trying to make that look big next to an Apollo launch?

We went to air launch, as I say, to get to the energy level that would get us to the 35-40,000 feet and up to launch speed without use of internal propellants. Incidentally, that's a marvelous way to go flying, as compared to the usual commotion of a takeoff with full power and pounding down runways. Airplanes aren't supposed to run fast on the ground, so that was a nice way to get flying. And that was a pretty nice airplane to launch. It didn't go out of the B-29, or the mothership, with as much negative acceleration as the X-1, and it came out flying pretty well. Generally you could get the engines lighting as you felt the shackles let go. We very seldom lost [much altitude] — oh, maybe 100 feet — and the airplane was on its way.

Also on the wing we tried a whole lot of devices to see if we could reduce the lateral transfer or flow of air that caused the separation of the air at the tips, which in turn caused the pitch-up and the tip stalls that were aggravated by the swept configuration. We put fences on it. We put more fences on the wings. We put notched leading edges. We put movable slats on the leading edge. And we put notches — different kinds of notches — different kinds of movable and immovable slats on the airplane. And really, not many of those things did an awful lot of good. The fences probably did as much good as anything, as I remember today. I do not believe that we put vortex generators on.

The technique would be to go up there and pull G at a fairly constant rate, trying to maintain as constant an air speed as possible. And incidentally, there was something we really re-learned with these kinds of wings. And that is that the old $C_{M_{C_L}}$[48] was a bunch of garbage as far as this goes. And we had to go back to the C_{m_a}. Because C_L was dropping so fast, that it looked like the airplane was going stable — when really it was going quite unstable at the time. And the airplane would pitch.

The worst pitching airplane that we ever flew, and we saw some of those problems, was the hard-wing F-86, which was just like this airplane when you had everything locked up, and had just a plain untreated wing. And, of course, they made that to get the speed to be the MiG killer that it was in Korea.[49] And if we could solve pitch-up, the techniques and methods we used with this airplane were those that became the design criteria in almost every design room that built swept-wing airplanes.

So that was a major contribution of the D-558-2, over a speed range of probably up to about Mach 1.5. We never really could do much when we got above those speeds. Because the speed wouldn't stay up there long enough to maneuver and

---

[48] $C_{M_{C_L}}$ represents the static stability in pitch of an aircraft. C_M is the pitching-moment coefficient. C_L is the lift coefficient. C_{m_a} is the partial derivative of the pitching-moment coefficient with respect to angle of attack — the angle of the airflow with respect to the wing of an aircraft.

[49] The F-86 averaged a 10:1 kill ratio over the MiG-15 in Korea, largely because the models used there featured the all-moveable horizontal stabilizer first flown on the XS-1 and the D-558s. Because their flight research was classified—although parts of the story about them were reported in the press—the Soviets were not aware of the benefits of the all-moveable horizontal stabilizer in transonic flight conditions.

accomplish what we wanted. And, of course, all of the work that we did with this airplane was to do it at enough altitude that the wing would stall before we reached its structural limits. I never had any of the roll-off with the D-558-2 that John and Bob Champine mentioned with this airplane. It was probably because I kept the ball in the center [laughter].[50]

You know, I had to be chuckling. I want to tell a little story I mentioned last night. Everybody up here has been having trouble with his memory, along with me. I couldn't remember what we did in this airplane. We were discussing that the other night at supper. And I mentioned that I was having a little trouble. I'd go into the flight service station. Couldn't remember whether I came in to close a flight plan or to open a flight plan. And Bob Champine says, "Yes. Same thing. I'd be at the top of the stairs, and couldn't remember whether I was going to go down or I'd just come up." Griffith says, "I don't have any of those kinds of problems at all. My memory is just as good as ever — knock on wood [sound of rapping]. . . . Come in!" And of course that guy Butch — Butchart's a frustrated fighter pilot. He had to fly torpeckers [slang for torpedo bombers] in the Navy.

Incidentally, the airplane you saw — that beat-up B-29 — you ought to see some pictures of airplanes that Stan Butchart brought aboard a carrier. They were unflyable. He violated the laws of physics; he brought some airplanes home that were pretty badly beat up. And that's one reason I gained quite a bit of respect for him. Because it wasn't to save the airplane. It was that I believe on two occasions he had a badly injured backseat man. And he wasn't going to abandon him.

Well anyway, Butchart checked me out in a B-29 — much as he claims I checked him out in a C-45. That was very interesting. And I'm going to make a long story short. I said, "Do you stall a thing like this?" I'm an old fighter pilot. Stalls are pretty common. He says, "Yeah, you're flying it." So I pulled it back. And it began shaking a little bit. And I looked over at Butch, and I said, "Do you go very deep into the stall?" And he said, "You're flying it." And the guys in the back were beginning to scream. Because things were shaking pretty badly. Well, I didn't want to seem to be chicken with this whole thing. So I looked at him, and he was very calm. Same way he is right now. So I pulled this thing back. And it was shaking. I'll tell you — I'd never been in a Tehachapi earthquake that was shaking so badly. All of a sudden, the right wing went out. And when those four engines started going around, you knew you had your hands full. Well, it was with consummate skill, I got this thing out of the spin into a screaming dive, then leveled out. I think probably I was shaking a little bit by then. I wasn't worried about me. I was worried about how I was going to explain to Vensel about that damn B-29. I looked over at Butch and I said, "What did you let me do that for?" He said, "You were flying it." That's his idea of a check-out.

Incidentally, these guys did me some good favors, too. Bob Champine went back to Virginia. So Griff hired me. And then Griff got an offer from Chance Vought and I said, "Hey, take it — take it." He left me with a fleet of the finest airplanes that a man ever got to fly. Howard Hughes couldn't afford the airplanes that I got paid to fly. And it was a good crew. And it got so there we had a pretty good flight rate for awhile, before we moved to the good laboratory up on the north end. And it would be like an X-1 to fly for breakfast, X-4 for lunch, and a D-558-2 in the afternoon. And

---

[50] A reference to centering the turn-and-bank indicator.

where could you get it better than that?

Of course the D-558-2 was one of our major projects. One of the things that very few people know is that when we were looking at a lot of the work we were going to do and then follow up with the X-2, I began working on what became the grandfather of all of our current full-pressure suits. The full-pressure suit final development was done at NACA at Edwards Air Force Base, and we went a long ways to doing that — to building the full-pressure suit.[51]

The first operational flight of a full-pressure suit was done by Marion Carl. He wore an exact duplicate of the one that we were developing. On the altitude flights, he went to 83,000 feet for the unofficial record that Butch mentioned. I mention that because it was kind of the way we did things in those days. Nobody ever said we could. But nobody ever said we couldn't. And we never asked permission.

The full-pressure suit — much of it was sewn on my wife's sewing machine. Incidentally, it was a Clark development. And very few people realize that David M. Clark probably was involved with the development of every bit of soft goods a pilot has worn since probably 1937 — whether it be G-suits, coveralls, helmets, ear protectors, much of the electronic gear, or that sort of thing. And he put a lot of his own money in it. And sadly, Dave Clark's gone. Because the nation misses men like him. He sent his people out here. And they lived at my house. And I built the back pad for the suit in my garage. We welded up the pressure bottle, and stress tested it in the shop. Jim Artz welded that up, I believe. Maybe it was Eddie Lane — names I think a lot of you people remember.

So we built this pressure suit. I built the console to test the suit, and did that sort of thing. The way we did things in those days is: I bought the regulator that had the gas ventilation flow go through it from the local gas company for seven dollars. The only bureaucratic problem I ran into was, when I wanted my seven bucks from NACA, I'd lost the receipt. And so I had to sign a voucher and say I really bought this thing. Now if you can imagine the United States government today allowing you to use something that only costs seven bucks, and didn't have any paper on it other than an invoice receipt, well, then you might begin to understand how it is we got some things done in those days, in that glorious era where everything that couldn't be done was done in flight test, based on professional judgment and just moving ahead.

It was, though, on these flights, as Butch described, that we began to realize that we ought to have only one guy on the radio. And he preferably ought to be a pilot. That finally went over through the development that we did on Apollo, and Mercury and Gemini, where they had an astronaut as the one guy on the radio. Because it would get so you had 15 people talking to you. The hydraulics guy wanted you to do this. The aerodynamics guy wanted you to do that. And finally, I had a habit of just turning the damn radio off, which didn't help my reputation with the people on the

---

[51] See documents 31-34 of this volume. By themselves, these documents are misleading and need to be read in conjunction with the narrative. The Air Force pressure suit mentioned in document 31 was not a full- but a partial-pressure suit. It was used extensively in flights at Edwards. The Navy full pressure suit discussed in documents 32-34 was much more developmental than the documents suggest. Scott Crossfield is emphatic that full credit for its development should go to Joe Ruseckas of the David Clark Co., who worked closely with him in the development effort. As Crossfield says in the narrative, much of the development took place at Edwards and in his garage at home.

ground very much. But it sure helped me get things sorted out.

On the D-558-2, I made the first NACA air launch in aircraft number 3, NACA 145, that used the rocket [as well as a jet] engine. And on that flight, we were going to go up and start exploring the transonic characteristics of setting the trim, and comparing it to the X-1, as we described here earlier, to see how much difference there was in the AC[52] shift of the swept wing versus the straight wing. And it was appreciable, in that it had a larger chord on the swept wing. And the percentages were about the same, but the moments were a great deal larger.

On that flight I lost an engine. The J34 engines did not like altitude, in spite of one of their representatives here today who claimed they did. I went through about 35,000 feet on the rockets. And the jet engine sounded like a .50 caliber machine gun going off. Very similar to the problems we ran into with a similarly constructed engine in the X-4s. I lost the engine. And as the engine was spooling down, I quickly lost all of the electrical power.

This was because of another totally isolated problem — that the reverse current relay wouldn't cut in until the generator output was down to 11 volts. And while it was coasting from what it needed — around 18 volts to 11 volts — I had no radios, no electrical power, no instruments. I also lost cabin pressurization and ventilation. And the windshields iced over. So about the only choice I had was to put the sun on a spot in the windshield, and then fly the airplane so it stayed there and so I knew at least the airplane was right side up. And it was doing something it was supposed to do.

You cannot fly blind. That's absolutely true. Your sense of balance, and your ears, and your eyes, and all of that, will not let you fly by the seat of the pants, blind. So that was really the only instrument I had at that time. The needle was beginning to wind down, and I didn't trust it, because it was electrically powered.

John Conrad came up on my wing, and he just told me what to do — lower the wing, raise the wing, and all that. He brought me all the way home. I owed him a drink, and I bought it for him. Incidentally, as Fitz Fulton reminded me the other night, he was the other chase pilot on that flight that day.

That was one experience with the D-558-2. And so from then on, we were very cautious. We didn't take the jet engine out to speeds at altitude. The reason for the problem on that was, it was one of the first engines Westinghouse made, or anybody made, that had an annular burner can.[53] And when you got way out of design pressure altitudes, the rotary component — the air going through the engine — would cause quite a radical increase in pressure. And the turbines weren't really seeing what the temperature was telling you or the loads on them were telling you. This was a common problem with those engines, until they put flow straighteners in them coming out of the burner can.

The way we did things then was something that I would like to leave, if I leave anything with this group here. Because if you remember — we went Mach 2 in 1953. Today the only airplanes that ever went significantly in excess of Mach 2 some

---

[52] AC is aerodynamic center — the point in the cross section of the wing about which the pitching moment stays practically constant despite changes in angle of attack. It is the center of lift with respect to the chord of the wing.

[53] An annular burner can was a combustion chamber on a gas turbine engine that had circular inner and outer boundaries. (Can was simply another name for a combustion chamber.)

40 years later are all in museums. There's something un-American about that. There's something we ought to be able to leave here that would encourage younger people to take the risks and the gambles. And I don't mean with personal hazard — I mean the technical risks and the monetary gambles that it took to get where we were going in those days.

I'd like to give an example of how we worked on those airplanes. We had a lakebed that looked like your lakebed looks out here today [that is, filled with water]. And we put a drogue 'chute on the airplane. I designed the drogue 'chute. Jim Artz welded up the piece. Then we riveted the container for it on the back of the airplane, and literally used parachute pins to open some spring-loaded doors. And that would pop out a little drogue that would pull the big drogue out.

We were going to have the capability to use the 5,000-foot strip down on the South Base,[54] which was all we had at that time — and keep flying when the lakebed was wet. Unfortunately, that same problem with the reverse current regulator came to bear again. On the first landing, I rolled down, the engine was spooling down, and, of course, the battery didn't cut in. There wasn't enough voltage to pull the pin on the parachute, so I rolled all the way down on the runway — and fortunately again, with consummate skill, managed to save the airplane with a half ground loop at the end of the runway — and then heard the thing come out. And the parachute fell on the ground! My reputation as an inventor didn't last very long after that sort of thing. We fixed the 'chute so it did work, but we never did use it on the runway.

Those were the kinds of things we did. The pilots had a big involvement and participation in what we did with the airplane. We used professional judgment. And we never had to ask anybody in Washington or the Air Force about what we wanted to do. A lot of times we weren't really sure what we were doing, except that we could make some plans of our own, based on professional judgment. There are many more stories like that and like the development of the pressure suit. And the reason for this long-winded dissertation is that I would hope the young people that are coming along now would say: "Hey, I can do that." And go do it. No more of this "whose budget is it going to come out of? Well, we tried that before. Did you think of this?" And all of those cop-outs that caused all of our failures to be in direct proportion to a reason or explanation of why we didn't do something.

With the 144 airplane [which had its turbojet engine replaced by a rocket engine in 1950], I did a dead-stick[55] landing. Picked up a Joshua tree on landing, and I got a little bit of ribbing from the crew. They photographed that Joshua tree that I'd picked up in the landing gear, and put the photo in a frame. And it hangs in my den today. So if my conceit needs calibrating, I can contemplate this and [laughter].

To get to the high-speed flights, I'd like to make reference to Bill Bridgeman. All of these airplanes had a characteristic that was called high-speed yawing or the instability that came with high speed. It really came with high speed and high altitude. The high altitude reduced the aerodynamic damping. So any small instabili-

---

[54] According to James O. Young, *Meeting the Challenge of Supersonic Flight* (Edwards AFB, CA: Air Force Flight Test Center History Office, 1997), p. 28, the main runway at South Base of Muroc Army Airfield (later renamed Edward Air Force Base) was already 6,500 feet long in October 1946. In any event, it was not exceptionally long.

[55] That is, without engine power.

ties were magnified to a large degree. Those things that did in Yeager, and Apt, and Murray on the X-1 airplanes and the X-2 airplanes, were also similar to characteristics on the D-558-2 — probably more like the X-2 because of the inherent dihedral[56] that we got from the swept-wing. And it would maybe oscillate once and then diverge. Or if you were at too low a G, it would diverge — directionally diverge [from straight and level flight].

Bill Bridgeman found that by manipulating the G, you could control the rate of this divergence, and give yourself time to get in very soft controls to hold it on almost a knife edge, if you please. He taught that to me. And by virtue of learning that, we overcame that "supersonic yaw," as the newspapers called it. And we managed to take the airplane out substantially beyond its expected design speeds. And in the course of that, we also were doing intermediate flights, going along with this stuff.

And we never did get into any of those instabilities that we were right on the ragged edge of all the time. And it was largely a flying technique. Because really these divergences and motions that we got into were not as expected as many of the other things that we encountered in high-speed flight. We knew that the X-2 was unstable directionally, statically, and dynamically at a certain speed.[57] We knew that the X-1A and D were unstable at a certain speed directionally, both dynamically and statically. But we really didn't know the manifestation of "supersonic yaw" and why it was happening in those days. In fact, do you remember why they had that huge tail on that X-15? I put that on there.[58] Fighter pilots need a lot of tail. And you don't have to live with those instabilities if you have an idea of what it is all about.

On the Mach 2 flight, people claim that I went to the Navy and got them to convince Dr. Dryden to let us do that. And I really didn't do that. I just dropped a hint to the Navy — that wouldn't it be great if they could whip Yeager's ass, and beat him to Mach 2! We knew that we had a very marginal situation. We were determined that if we did everything just right, we could thread that needle right out there where Ankenbruck had calculated it, and come out with about Mach 2.03, or something like that.[59]

---

[56] Inclination to roll about the longitudinal axis.

[57] Static stability is the ability of an airplane to return to straight and level flight after it has been disturbed by an outside force, such as atmospheric turbulence. To give one example, if the turbulence forces the nose up slightly, a statically stable aircraft will return to level flight. If the aircraft is statically unstable, on the other hand, it will nose up still more than it did initially. Dynamic stability is the property of an aircraft that enables it gradually to reduce an oscillatory motion produced by an outside force and return to straight and level flight. The aircraft is dynamically unstable if it increases the magnitude of its oscillations unless controlled by the pilot.

[58] To provide some background to this development, analytical studies at the NACA's Langley Aeronautical Laboratory by a team headed by John V. Becker indicated the need for a large cruciform tail configuration on the X-15, with a "wedge" vertical fin to give an increase in effective vertical fin area.

[59] Herman O. Ankenbruck was the project engineer on the D-558-2 who designed the flight plan to achieve Mach 2 by climbing to about 72,000 feet and pushing over into a slight dive. Hallion, *Supersonic Flight*, p. 179.

We put nozzle extensions on the propulsion system. I had earlier on put the tank regulators up in the cockpit, so that once we started the engines with the normal pump inlet pressures, I could crank up the tank regulators another 10 or 15 pounds. That was magnified with a pump inlet pressure. And that would give us a lot higher pressures in the rocket — in each of four rocket chambers. That, with the nozzle extensions, gave us an airplane that almost had 9,000 pounds of thrust, as compared to the spec 6,000 pounds of thrust. It also burned the fuel a lot faster — appreciably faster, anyway.

So after launch, and I got the four engines going, I would crank up the regulators. It was just a little bit of a throttle to get them up. The nozzle extensions, I think, gave us probably 1,000 of the couple thousand pounds of thrust that we gained. It wasn't the first time we'd used them. But it's the first time we ever used them to total advantage.

We took that airplane. And everybody said supersonic parasite drag is not a thing of consequence. But we didn't give a darn. We taped every crack in that airplane. We polished it. And it just shone like a — I won't say it. It would get me in trouble politically. And it was very smooth. We took every bump off of it, and sanded it.

And then one of the things we did — we were looking to remove every pound of drag we could on the airplane. The two jettison lines that stuck out the aft end of the airplane that kept the fuel away from the B-29 were not really an essential part of the D-558-2 if we launched. So Jack Russell made a couple of aluminum lines. Instead of going straight out to jettison this, they curved back into the rocket engine wake. So if I launched and fired the engines, they burned off and fell on somebody's house out here. But we didn't have to carry them around with the additional drag they produced. There was also another overboard vent line that came from the fuel tank off the side of the airplane. We took that off the airplane, and put it on a bracket on the B-29 so that when I dropped away, that was flush. So there just was nothing sticking out on this airplane anywhere.

On the night before the flight, we cold soaked the alcohol all night long using a big refrigeration unit. We got it so cold that probably we added another 10-15 gallons to the capacity of the alcohol tank. And we also cold soaked the airplane. We loaded the liquid oxygen (lox) in it very early that night before, and then kept upgrading it all the time so that instead of being at -292 degrees, it was probably colder than that. And we got more lox on board.

That almost did us in. The next morning, it was so cold that when they loaded the peroxide, one of the overboard vents choked with ice. And so the pressure from the loading peroxide vented it out through another part of a manifolded venting system for that tank and sprayed out of the airplane and on to one of the mechanics — Jack Moise. Jack hollered, and put his hands over his face. And another chap named Kincaid — and I can't think of his first name. Do you remember, Vicki?

VICKI IKLER [a retiree from the audience]: Gil.

CROSSFIELD: Gil? Oh, that's a drink measure, isn't it? Okay. Now I know. I'll never forget it. Anyway, he grabbed the hose. And he hosed down Jack Moise. And we immediately bundled him over to the nurses' station. Jack went inside. And the nurse began working on his face, washing out his eyes, and that sort of thing. We were concerned. Because peroxide is a very vicious chemical — very active. Well, I took him over there. I saw that Jack was being taken care of, and I was going to head out for the airplane.

Then I looked over at Kincaid, and he was sitting there soaking wet. And it was a bitterly cold morning. I said, "You must be cold. Aren't you freezing to death?" He said, "No. I'm really quite warm." "Oh, that's great." Whoa! — I turned around. He was cooking. He was full of hydrogen peroxide himself. And he was getting warmed up from that. He thought I was nuts because I began pulling his pants off, and he didn't know what my intentions were. Well anyway, he had two pairs of pants, and two pairs of winter underwear on. When we got down to his legs, he had those white spots on him that were characteristic of hydrogen peroxide burns.

So that was the kind of morning we started out with. I had a pretty bad case of the flu. But I wasn't about to give up after all the work that crew had done on that airplane. We went out, and very fortunately, we had a little help from wind shear and a flight plan that worked for a change. I was up on the edge of my seat. Everybody was that way. And that day I didn't even have to turn off the radio. So I had a lot of advice, which was a bad habit. I went out and threaded that thing, and we made our Mach 2.001 — or 2.005 they said. It was 1,338 mph. Then they finally re-corrected it to 1,291 and something miles an hour. Mach 2.005, I believe, was what they gave me. That sounds like a little press release because I made it past two. Walt was pretty happy. And we were happy. It means nothing technically. It meant nothing from a research standpoint. It only meant that we got in the ball game, and we got a score on the board. And we beat Yeager there that time.

The Navy was pretty happy with that, and made quite a bit of a to-do about it. And I was pretty happy, because I was invited to the 50th anniversary of flight down at San Diego where the Ryan Corporation was celebrating. And my dinner partner that night up on the head table was Esther Williams. So see, there are rewards for. . . .

The sequel to that story is it almost caused me a divorce. Esther got up to make her speech. And she said, "You know, I've been getting a lot of static all night long about sitting next to the fastest man on earth. But I don't believe it. He hasn't laid a hand on me yet!" So without thinking — or maybe I was thinking — I reached over and swatted her on her beautiful behind. And my wife never did forgive me for that.[60] Thank you.

HALLION: I think we've had a really great day. We're running just a little bit behind. But with our panelists here, we'll take at least ten minutes Q and A [questions and answers]. So gentlemen, and audience, the floor is now yours. I'll repeat the questions, for those who may have trouble hearing them. Do we have a question out there?

DILL HUNLEY: Dick, this isn't just about the D-558. But the two D-558s and the X-1 shared the movable horizontal stabilizer. And there was a video the British put out last year that attributed that innovation to British research. Do you know if there's any truth to that?

---

[60] This story, including the quotation but not some of the details, is also told in Crossfield's *Always Another Dawn*, p. 179.

HALLION: It's an utter myth. This obviously has been a symposium concentrating on the D-558. But there is something here that we have to talk about regarding the X-1 for a minute. The British television program was picked up by *Nova* [a Public Television Broadcasting Station television series in the U.S.A.] — and to give *Nova* its credit, it recognized there were a lot of flaws in it. The *Nova* people tried to work as much as they could with the video. They had to work around those flaws and some of the problems in it. The video that had been done in Great Britain suggested that there was a technology transfer from the Miles Aircraft Corporation and the so-called Miles M.52 program, which was a proposed transonic research airplane that never went anywhere.[61]

The video suggests that there was a transfer from the Miles M.52 effort into the XS-1. Absolutely false. Partisans for the Miles M.52 program suggest that after it was canceled, data was transferred to Bell. At the time that the M.52 was canceled, the XS-1 was already flying. Its design had already been fixed. And there was no possible way that there could be any technology transfer there.

This same issue on the all-moving tail — you know, actually the all-moving tail, if we think about it, is like other devices that we've experienced — the flying wing, the swept wing, for example. You know, if we go back in time, you can find predecessors. But you have to think: why were the people actually applying this technology to a particular aircraft design? In the case of the swept wing, which we have talked about today as a means of alleviating transonic drag rise and achieving good high speed performance, the concept actually dates back to the days of John Dunne, before the First World War, where he was using the swept wing to alleviate stability and control problems with tailless airplanes. It bore no relationship to the high-speed requirements that people were looking at in the 1940s.

We had actually had all-moving tails appear as early as pre-World War I airplanes. In fact, if you take a look at the Wright flyer, you have here a canard surface that's an all-moving surface. But that's a very different thing from what people were actually trying to do here.

The NACA at Langley field in the 1940s undertook some very interesting research with an airplane called the Curtiss XP-42, which they modified to have an all-moving tail. And they studied the benefits and the advantages of the all-moving tail thoroughly and recognized, certainly by mid-1944, that if you were to develop a high speed research airplane, that would be a very desirable attribute for the aircraft. So this idea that somehow the all-moving tail we've talked about was something that we gained because we had exposure to or benefited from some foreign research — once again, that's simply not the case. And I'm glad you raised the question, Dill, because it's an important point to bring out. Other question here?

NEW SPEAKER (unidentified): I'll throw this to whoever wants to catch it. I understand a lot of the testing that's gone on. You did your structural testings and your coupling and rolling. But I heard no mention of the structural aerodynamic aero-elasticity validation of the airplane, and was just wondering what particular techniques you used to get stabilized on a dynamics point.

---

[61] It was canceled in 1946 according to *Jane's Encyclopedia of Aviation*, Michael J. H. Taylor, ed. (rev. ed.; New York: Crescent Books, 1996), p. 675.

HALLION: Okay, Scott. It's all yours.

CROSSFIELD: That one's kind of easy to answer. Number one, the structural demonstration was the responsibility of the contractor. And Douglas demonstrated the airplane would meet the design-limit stress, and took it just beyond limit to establish that. But these airplanes were built so strong, that they were — aerodynamically for what we were doing — virtually rigid. And so the aero-elastic effects hardly ever showed up in the airplanes. The natural frequencies of the wings were very high. And they didn't have much effect on the kind of stability-and-control handling qualities that we're talking about. And that's one of the reasons it didn't show up.

On the D-558-1, there was a rudder buzz at about Mach 0.999, just as it was going to Mach 1.0. And I don't believe anybody ever went into that rudder buzz. That was the only dynamic problem that I remember on any of those airplanes. But it's primarily the strength of the airplanes. They were 18 G and 12 G airplanes, respectively [that is, 18 Gs for the D-558-1 and 12 Gs for the D-558-2] — very rigid.

HALLION: Okay, Scotty. Next question?

NEW SPEAKER: Dick — Where did the designation 558 come from?

HALLION: The Douglas D-558 designation was a company designation. Douglas used that prefix and numbering system for its own aircraft. You know, it's really funny. Because when Ken [Szalai] was getting the symposium together, there was this idea of calling the symposium "The X-Planes That Weren't." And it's really true, you know. If we think about it, these were X-airplanes — undoubtedly. But they were just like the XF-92A, which ostensibly from that designation, you'd think was a prototype fighter, but in point of fact, it was a delta-wing technology test bed, was an X-airplane. But the X designations, as they started out, actually were XS designations in those very early days. And they were basically the province of the Army Air Forces — later the United States Air Force. And it was not really until we got beyond the X-15 era that we started thinking of the X designation as a national designation system, so that it was applied to aircraft that came from organizations other than the United States Air Force. And remember the X-15 had a three-man executive steering committee, beyond the NACA research airplane projects panel. You had an Air Force, a Navy, and an NACA — later NASA — representative steering that. So that clearly, you know, if we were developing the D-558-1 and -2 today, each one of them would undoubtedly have a separate X-series designation. But the D-558 was a corporate designation.

There was, incidentally, a D-558 that we haven't mentioned here today. And before the conference concludes, we should mention it. The Office of Naval Research was very interested in hypersonic flight. And in response to that, Heinemann and his design team put together a proposal for a so-called D-558-3, which would have been a Mach 6 research vehicle. And that was one of the concepts proposed for what eventually became a competition among several aircraft manufacturers that resulted in the North American X-15. But that was the D-558 that never was, so to speak.

Other questions? Well, I think we've had a very good session. And I'm sure there're a lot of you who want to meet in person with our panel. I want to thank you all for your attendance today. And I want to thank the leadership of the Dryden Flight Research Center for having put this program together. I think Ken Szalai is out here in the audience someplace — or he was earlier. And Ken, you very much deserve a kudo for this. I must say, the activity of this Center over the last few years in putting together historical symposia has been outstanding. So, here's one for you. Thanks a lot. [Audience applause]

Cover of *Flights of Discovery* showing William S. Phillips' painting *Mach 2 Dawn* from the NASA art program. The painting depicts Scott Crossfield's Mach 2 flight in the D-558-2 described in these pages.

SCHNEIDER: Well, I was thinking, first of all, of how fortunate we are to have Dr. Richard Hallion — not only as a historian, but as a biographer and as an aviation advocate. Secondly, how fortunate this country has been to have these pilots. Also, the design team at Douglas, and the crews that made the airplane what it was, and created this great database that this country has built its supersonic and transonic capability on. Third, how fortunate we are to have these people here today, so many years later, and able to talk to us personally where we become a part of this history now by participating in it. And finally, how fortunate we will be if we really listen carefully to the lessons learned, think about them, and apply them in our own areas of responsibility.

I want to close this up by reading off the names of all the pilots that flew the D-558. And I'm going to group them together. And I'll identify them by organization:

From the United States Marine Corps: Maj. Gen. Marion Carl. From the United States Navy: Captain Frederick Trapnell and Commander Turner Caldwell. From the United States Air Force: Lieutenant Colonel Frank Everest, Major General Al Boyd. From the Douglas Corporation — these are gentlemen that really had a tremendous number of flights in the airplane: John Martin, Eugene May, Bill Bridgeman. And lastly, from NACA: Bob Champine, Howard C. "Tick" Lilly, John Griffith, Scott Crossfield, Walter Jones, Stan Butchart, Joe Walker, and John McKay.

I want to thank these participants. And we're very pleased to have a representative, Charlie Delavan from Douglas, to help celebrate this great anniversary today.

And in recognition of that, we have a small memento. We'd like to ask the four pilots, Charlie Delavan, and Dick Hallion please to come up here. And I'll present a small token of this day. Just come on up here.

We have a copy of a painting for each of you — *Mach 2 Dawn* — a very famous painting, which also happens to be on the cover of our 50-year history, *Flights of Discovery*. And we're pleased to present each of you with this as a remembrance of this day. So we'll start here.

[PRESENTS COPIES]

So thank you, gentlemen. [Audience applause]

[END OF SYMPOSIUM]

Appendix — The Aircraft

Douglas D-558-1

The Skystreaks were roughly 35 feet long, 12 feet high, and 25 feet across the (straight) wing span. They were powered by one Allison J35-A-11 engine (developed by General Electric as the TG-180), which was rated at 5,000 pounds of static thrust. The airplane carried 230 gallons of aviation fuel (kerosene).

NACA 140 is located at the Naval Aviation Museum in Pensacola, Florida. NACA 142 is at the Marine Corps Air Ground Museum, Quantico, Virginia.

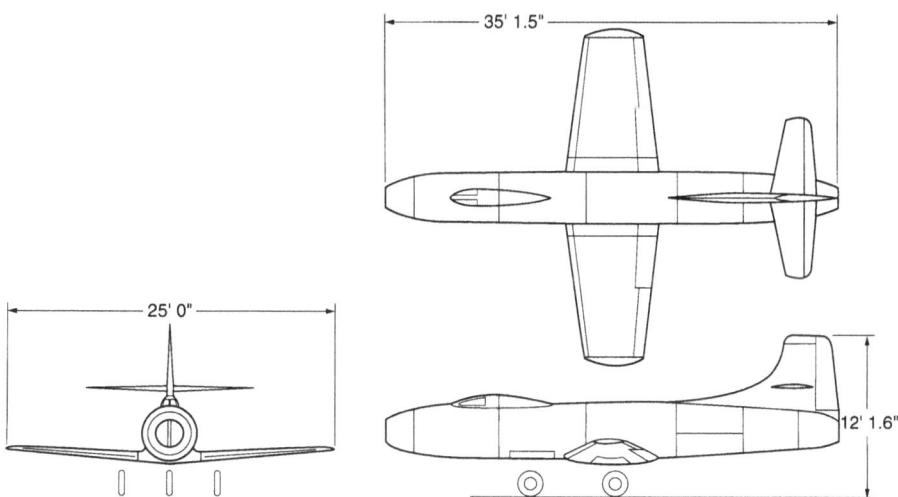

Douglas D-558-2

All three of the Skyrockets had a height of 12 feet 8 inches, a length of 42 feet, and 35-degree swept wings with a span of 25 feet.

Until configured for air launch, NACA 143 featured a Westinghouse J34-40 turbojet engine rated at 3,000 pounds of static thrust. It carried 260 gallons of aviation gasoline and weighed 10,572 pounds at take-off.

NACA 144 (and NACA 143 after modifications in 1955) was powered by an LR-8-RM-6 rocket engine rated at 6,000 pounds of static thrust. Its propellants were 345 gallons of liquid oxygen and 378 gallons of diluted ethyl alcohol. In its launch configuration, it weighed 15,787 pounds.

NACA 145 had both an LR-8-RM-5 rocket engine rated at 6,000 pounds of thrust and a Westinghouse J34-40 turbojet engine rated at 3,000 pounds of static thrust. It carried 170 gallons of liquid oxygen, 192 gallons of diluted ethyl alcohol, and 260 gallons of aviation gasoline for a launch weight of 15,266 pounds.

NACA 143 is currently in storage at the Planes of Fame Museum, Ontario, California. The second Skyrocket, NACA 144, is in the Smithsonian Institution's National Air and Space Museum in Washington, D.C. NACA 145 is on display in front of the Antelope Valley College in Lancaster, California.

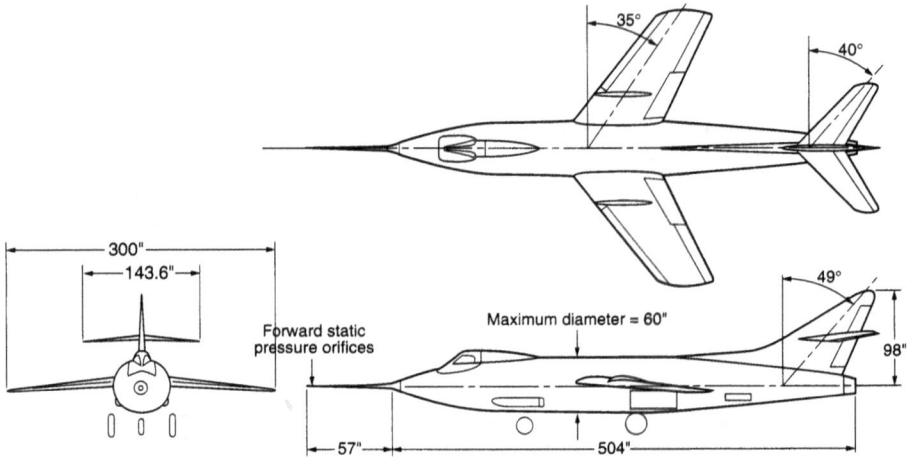

Documents

CONFIDENTIAL

Langley Air Force Base, Va.
June 13, 1949

From Langley
To NACA

Subject: Discussion of D-558-1 airplane projects at NACA Headquarters on June 8, 1949

1. An informal discussion was held in NACA Headquarters on June 8, 1949 to acquaint the Bureau of Aeronautics personnel with the status of the D-558 phase 1 and 2 projects and with other problems at Muroc that are related to these projects.

2. The discussion was attended by:

        H. A. Soulé
        C. H. Helms
        M. N. Gough
        R. O. Robinson
        J. W. Crowley, Jr.

        W. Z. Frisbie
        C. W. Stirling, Comdr., USN
        S. W. Kerkering, Lt. Comdr., USN
        F. C. Riley, Lt. Comdr., USN
        O. Bessio
        R. A. Carl, DE
        G. O. Kayten, DE
        F. A. Louden, DE
        W. S. Diehl, Capt., USN

        R and D

        Lee R. Standifer, Capt., USAF

Mr. Soulé opened the discussion with a review of the D-558-1 airplane project. He dealt with the flights made this spring on airplane No. 37972, the instrumentation, the present plans for this airplane, and the condition of airplane No. 37973. He also mentioned the reports that have been issued or are being prepared as a result of the flights to date. After a short period where questions were asked regarding the D-558-1 project, he presented similar information for the D-558-2 airplane project.

CONFIDENTIAL

Document 1: Memo, Hartley A. Soulé, [NACA] Research Airplane Projects Leader, To NACA, Subject: Discussion of D-558-1 airplane projects at NACA Headquarters on June 8, 1949, Date: June 13, 1949

3. In the discussion of the D-558-1 project, reference was made to the Bureau letter to the NACA requesting that airplane No. 37973 be maintained so that it can be flown after release by the NACA. In this connection, it was reported that the airplane is being maintained but that the tail was currently at Langley being instrumented for aeroelastic measurements of elevator twist. On completion of the installation of the instrumentation of this tail, it will be traded with the tail on airplane No. 37972. It was also reported that the NACA considered the procurement of pressure distribution on the phase 1 airplane a very important project and that the Committee desired to complete the measurements planned. It was estimated that not more than 25 percent of the work planned for the airplane had been completed. Reference was made to the fact that the tests may proceed slowly because of the difficulty of maintaining the airplane. In particular, the instruments have to be removed each time an engine inspection is made.

4. In a discussion of the D-558-2 project, the question was raised regarding the return of the airplane to the Douglas Company for the installation of the rocket engine in airplane No. 37974. In reply, it was stated that the NACA had just completed the installation of instruments and had started on the initial flight program for the airplane. It was extremely desirable that the NACA be permitted to complete the work on this phase before the airplane is turned back to the Douglas Company. The program may be completed before the end of September, but the airplane will, in any event, be returned to the Douglas Company in December. It was reported that the Douglas Company personnel had contacted the Muroc Unit with regard to the possibility of air launching the D-558-2. The Muroc Unit supplied information to the Douglas Company as requested. The Douglas personnel stated that the Douglas Company would make a study of the costs of modifying a B-29 and D-558 phase 2 airplane for air launching, and would approach the Bureau with a definite bid for the work. With regard to the air launching of the phase 2 airplane, it appears desirable to eliminate the jet engine for the test. There is no need to hold up the present program while an airplane is being modified for air launching.

5. With regard to related situations at Muroc, Mr. Soulé reported that there are two pilots assigned to the Muroc Unit. As an aid to the operations at Muroc, the Bureau had previously offered the NACA a starting unit for jet engines. This unit has not arrived, and the Bureau representatives agreed to check as to the availability of the unit. The housing situation is critical at Muroc. The Bureau has supplied houses at Mojave Air Station. The number of houses being utilized, however, is greater than the number assigned. The situation is becoming more critical, as the staff is expanded to accelerate the work on the research airplanes and it appears desirable to have the present position of the Committee at Mojave regularized and, if possible, more houses assigned for further expansion. In this connection, reference was made to a suggestion of Commander Vjetasa that the Bureau open up and modify as apartment houses some of the 200-man barracks at the field that are not being utilized. The

possibility of obtaining an airplane for Muroc that would not be so experimental in nature as the research airplanes so that more flights could be obtained by the pilots at Muroc to keep up their pilot proficiency was discussed. The Bureau representatives indicated that it might be possible to assign a McDonnell F-2H airplane to Muroc for a flying-qualities investigation. They agreed to check into this possibility.

                                              Hartley A. Soulé
                              Research Airplane Projects Leader

HAS.cbm

FLT

CONFIDENTIAL

DECLASSIFIED
Authority NND 927631

CONFIDENTIAL

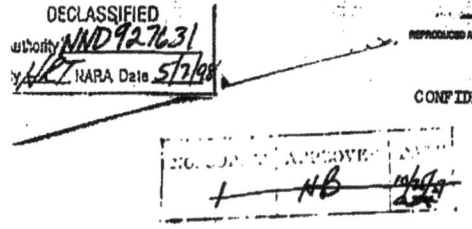

Muroc, California
September 19, 1951

MEMORANDUM to Chief of Research

Subject: Information concerning elevator vibration of the D-558-I airplane.

Reference: Letter from NACA to Langley, August 31, 1951, HHB/mb.

In the reference letter and also in a telephone conversation with Mr. Hartley A. Soulé a request was made for more detailed information concerning the torsional elevator flutter that has been encountered during flights with the D-558-I airplane. A description of the Douglas Aircraft Company's encounter with the problem along with time histories and other data from NACA flights during which the vibrations occurred, have been collected and are submitted below.

History of the Elevator Vibration.- The vibrations were first encountered during the Douglas demonstration flights No. 85 and 86 with the 37970 airplane. They occurred at the end of dives to Mach numbers of about 0.92 and during the pull-outs. The vibrations were asymmetrical and had a frequency of about 30 cycles per second. The mode of the vibration was such that a node exists at the fuselage line, hence, the damper attached at that point is ineffective.

In an attempt to evaluate the magnitude of the vibration the Douglas Company placed some accelerometers and twist indicators on the elevators which indicated a twisting of the elevator of about $\pm 2°$ and a balance weight acceleration of about 30g. Comparison with earlier, more violent, flights led them to believe that they might have reached three times these figures, although the balance weights were only designed for 60g. As a result of these measurements about one-third of the mass of the outboard balance weights was removed and added to the inboard weights on the 37970 airplane only. The results of these changes were never checked by the Douglas Company as no more high speed flights were made before turning the airplane over to the NACA. One of their engineers frankly stated that the move was a "shot-in-the-dark" and would probably have negligible effect.

The vibrations probably occurred on many of the early NACA flights with the D-558-I airplane but went undetected because of a lack of suitable instrumentation. However, during the summer of 1950 the complete horizontal tail from the 37970 airplane which is the one having the modified balance weights was placed on the 37972 airplane and adequately instrumented so that the vibrations could be detected. This assembly was used for flights 28 through 34 and the elevator vibrations were frequently encountered.

Because of the possibility of a fatigue failure resulting from the elevator vibrations and since the 37970 tail assembly had been used for more than 100 flights while the 37972 tail assembly had been used for only 25 flights, it was decided to put the original elevators and the unmodified balance weights back on the 37972 airplane. This configuration has been used for all flights from No. 35 to the present time.

CONFIDENTIAL

Elevator Vibration Data.- Figure one shows a 1/10 scale drawing of the right side of the horizontal stabilizer and elevator. The location of the various parts of the instrumentation are indicated. For most flights the elevator CPT's had adequate spring tension to follow a 30 cycle per second vibration. Originally all flat type CPTs were used but after flight 34 the outboard ones were changed to the pencil type. At first the pencil type CPTs did not have sufficient tension to follow the motion, but later heavier springs were installed.

Figure two shows the buffet boundary on a plot of airplane normal force coefficient versus Mach number and also plotted is the location of flight test data for which time histories are presented. The time histories are presented in figures 3 through 9. Each figure has tracings of the elevator position records and the stabilizer bending moment, shear, and twist records. Unless otherwise noted the conditions on the right stabilizer and elevator are shown. In addition, the Mach number, normal-force coefficient, and angle-of-attack is plotted on each time history.

Figures 3, 4, and 5 show data at altitudes of 15,000, 25,000, and 35,000 feet for comparable Mach numbers and normal-force coefficients. It might be suspected that the vibrations are more severe at the lower altitudes, but the variations obtainable at any one altitude far exceed the other variations. Comparing figures 4 and 6 which are at comparable altitudes and Mach numbers it can be seen that when the plane is held at the point of maximum lift and the angle-of-attack allowed to increase the magnitude of the fluctuations increases rapidly. During the run shown in figure 6 the maximum twisting of the elevator between the two CPTs was 5.3°. According to the Douglas Aircraft engineer this corresponds to about 70 percent of the design stress.

The effect of the change in the balance weights is shown in figures 7 and 8. The run shown in figure 7 was made with the modified balance weights and the run shown in figure 8 was made with the unmodified balance weights. There is no significant difference.

Figure 9 shows a run at low lift conditions yet at sufficiently high Mach number to be in the buffet region. It is apparent from this figure and the other figures covering a range of Mach numbers that there is no appreciable variation with Mach number. The greatest magnitude vibrations, thus far encountered, have occurred during stalls at comparatively low Mach numbers.

Summary of Results

1. The vibrations occur primarily under high lift conditions at all Mach numbers.

2. The vibrations are always associated with buffeting.

CONFIDENTIAL

CONFIDENTIAL

- 3 -

3. The amplitude of the vibrations is greatly increased by allowing the angle-of-attack to increase after maximum lift is reached.

4. The change in balance weights made by the Douglas Company had little or no effect on the vibrations.

Donald R. Bellman
Aeronautical Research Scientist

DRB:mem

cc: NACA Headquarters
Mr. Arthur Regier, Langley
J. Vensel, Muroc

CONFIDENTIAL

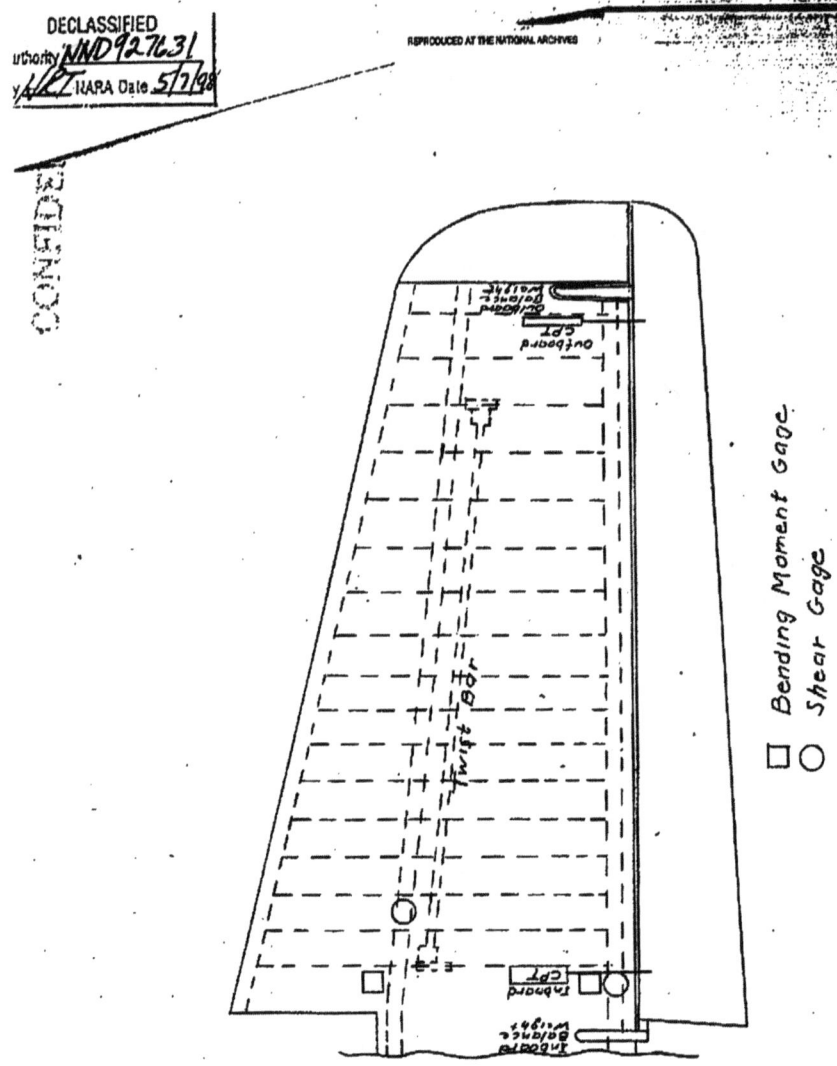

Figure 1. Location of strain gages and elevator CPT's Horizontal Tail - D-558-I, 37972 Airplane

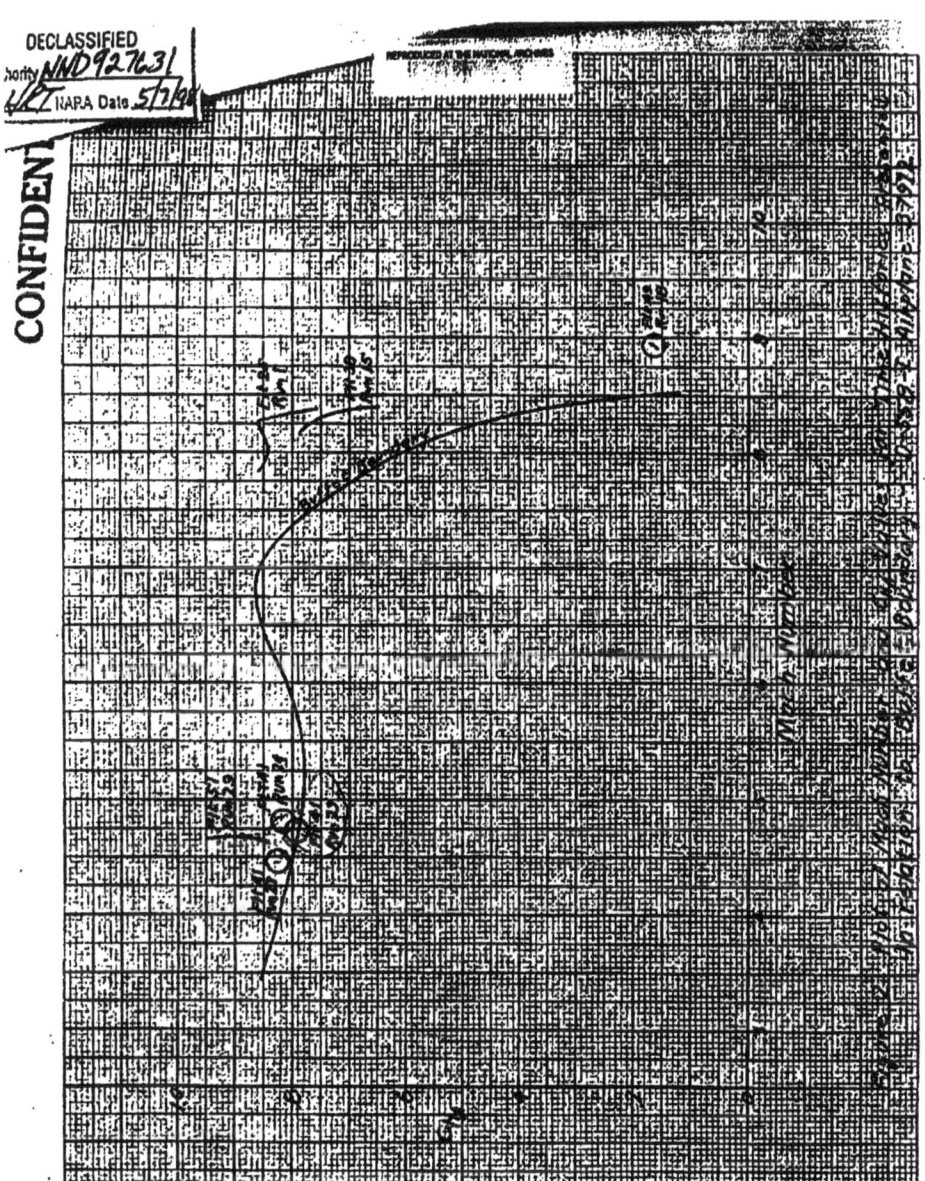

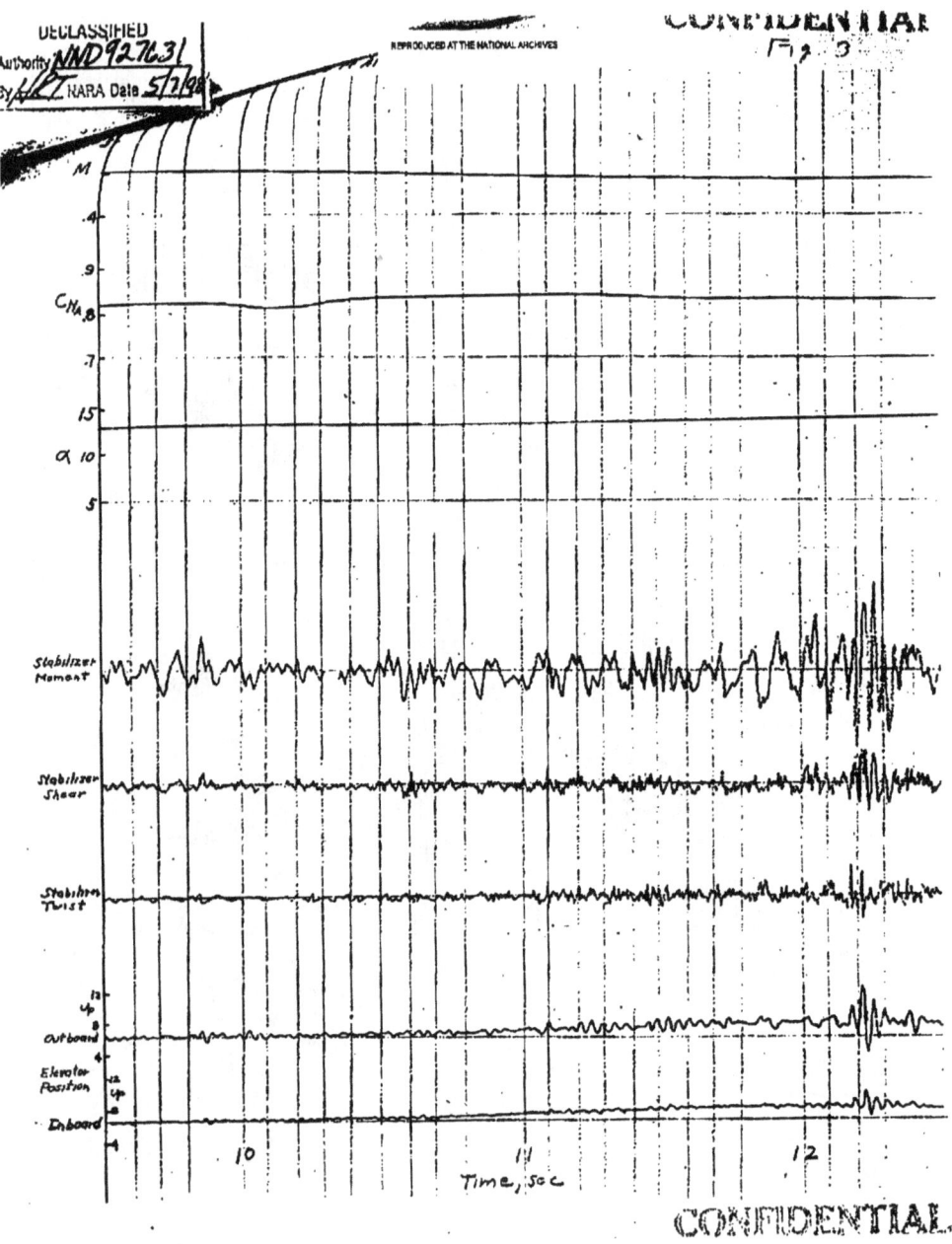

CONFIDENTIAL

Muroc, California
October 12, 1951

MEMORANDUM for Chief of Research

Subject: Progress report for D-558-I airplane (142) for the period September 22 to October 5, 1951.

1. The airplane was out-of-commission for the entire report period because of an inspection and overhaul of the stabilizer actuating mechanism.

2. In connection with the tail buffeting intensity variation with altitude, lift, and angle-of-attack, additional data was deemed necessary and is being worked-up.

3. The work-up of tail load data over the Mach number range from 0.47 to 0.88 has been completed except for pitching acceleration corrections which are required for some runs.

Donald R. Bellman
Donald R. Bellman
Aeronautical Research Scientist

DRB:mem

cc: NACA Headquarters (2)
    Ames
    Lewis
    ChBuAero
    Navy Liaison, Muroc
    Projects Engineer
    Files

CONFIDENTIAL

Edwards, California
July 23, 1952

MEMORANDUM for Chief of Research

Subject: Progress report for D-558-I (142) research
airplane for the period June 28 to
July 11, 1952.

1. Flight 58 was made on July 2, 1952 for the purpose of obtaining data in connection with the aileron effectiveness investigation.

2. Aileron roll data has been completed at 10,000 feet altitude for Mach numbers from 0.4 to 0.8 and at 25,000 feet altitude for Mach numbers from 0.45 to 0.85. This data extends and substantiates earlier roll data. Data up to the limiting speed of the airplane will be obtained at 35,000 feet altitude on subsequent flights.

Donald R. Bellman
Aeronautical Research Scientist

DRB:mem

DEB

cc: NACA Headquarters (2)
Ames
Lewis
ChBuAero
Navy Liaison, Edwards
Projects Engineer
Files

Edwards, California
July 30, 1952

MEMORANDUM for Chief of Research

Subject: Progress report for D-558-I (142) research
airplane for the period July 12 to
July 25, 1952.

    1. Flights 59 and 60 were made on July 17 and 22, respectively. Both were research flights for the purpose of obtaining data in connection with the aileron effectiveness and the dynamic longitudinal stability programs.

    2. Aileron rolls at the limiting Mach number of the airplane are still needed to confirm and extend the indicated decrease in the aileron effectiveness.

    3. The tail buffeting magnitude data has been extended and is being reviewed by the section head.

    4. The horizontal tail load data has remained dormant because the engineer is needed on the D-558-II tail load program.

                            Donald R. Bellman
DRB:mem   Aeronautical Research Scientist
DEB
WCW

cc: NACA Headquarters (2)
    Lewis
    Ames
    ChBuAero
    Navy Liaison, Edwards
    Projects Engineer
    Files

Document 5, Memo, Donald R. Bellman, Aeronautical Research Scientist, To: Chief of Research, Subject: Progress report for the D-558-1 (142) research airplane for the period July 12 to July 25, 1952, Date: July 30, 1952

Edwards, California
December 11, 1952

MEMORANDUM for Chief of Research

Subject: Progress report for D-558-I (142) research airplane for the period November 1 to December 1, 1952.

1. The airplane was returned to flying status in this period but heavy rains and wet lake bed conditions have prevented additional flights.

2. During the past year the lateral control effectiveness investigation was completed and the dynamic stability investigation was started. The flight maneuvers for the longitudinal and lateral dynamic stability program consist of elevator and rudder impulses. The data are to be obtained over a range of Mach numbers to 0.90 and at altitudes of 25,000 and 35,000 feet. At the present time about one-third of the elevator impulses and a few of the rudder impulses have been obtained. The data from the three most recent flights is being placed on IBM cards which will be sent to Langley for computation on IBM machines. Six to eight flights will be required for the completion of the flight investigation.

3. The tail load and tail buffeting magnitude programs have complete sets of flight data and partially completed analyses. Both programs are dormant because the engineers have been needed for work on the D-558-II and the X-3 programs.

                                                  for Donald R. Bellman
                                                  Aeronautical Research Scientist

DRB:mem

DEB
WCW

cc: NACA Headquarters (2)
    Lewis
    Ames
    ChBuAero
    Navy Liaison, Edwards
    Projects Engineer
    Files

CONFIDENTIAL

Muroc, California
September 13, 1951

MEMORANDUM for Chief of Research

Subject: Progress report for D-558-II (37974) research airplane for the period August 25 to September 7, 1951.

1. The airplane along with BuAero No. 37973 was delivered on August 31. The NACA designation for 37974 will be No. 144.

2. The Operations Section is inspecting the airplane, and work has been started on the instrument changes. The major instrument changes are installation of fuselage and base pressure orifices and manometers, and the connection of certain wing, horizontal and vertical tail strain gages.

3. It is expected that the NACA exploratory flight test program will begin in about two weeks.

Herman O. Ankenbruck
Aeronautical Engineer

HOA:mem

cc: NACA Headquarters (2)
Ames
Lewis
ChBuAero
Navy Liaison, Muroc
Projects Engineer
Files

Muroc, California
October 16, 1951

MEMORANDUM for Chief of Research

Subject: Progress report for D-558-II (144) research airplane for the period September 22 to October 5, 1951.

Present status

1. A flight was made as scheduled on September 28. The operation of the engine was rough on 3 cylinders. The inverter failed about halfway through the flight and some data was lost. The maximum Mach number reached was 1.2 at 45,000 feet altitude.

2. A preliminary analysis of the limited amount of data obtained to date has shown that:

   (a). The directional stability and damping decrease markedly as Mach number is increased to 1.85.

   (b). There are large changes in the rudder hinge-moment parameter, C_{h_α}, with power becoming positive at Mach numbers near 1.8.

   (c). A directional trim change appears to exist at Mach numbers near 1.4. It is felt that the development of the lateral oscillation is associated with this trim change, the changes in rudder C_{h_α}, and the fact that the rudder was not rigidly fixed. (Static tests showed that for hinge moments measured in flight it was possible for the rudder to deflect as much as $2°$ in the locked position).

   (d). There appears to be no wing dropping on No. 144 airplane at Mach numbers from 0.9 to 1.0. Data obtained with the No. 145 airplane shows that some initial aileron trim is required throughout the speed range, and that a slight wing-dropping occurred at a Mach number of 0.96. This wing dropping may in part be due to a measured decrease in aileron effectiveness at about the same Mach number. On the No. 144 airplane, some aileron deflection was used to correct the directional trim change at a Mach number of 1.4, noted in (c) above.

   (e). The data thus far obtained show nose down trim changes near a Mach number of 1.0 and a reversal to nose up trim near a Mach number of 1.1. This nose up trim change appears to reduce as the Mach number increases further to 1.8. As Mach number is increased from 0.6 to 1.7, the stabilizer effectiveness appears to decrease to approximately 30 percent of the low speed value.

There is no data between Mach numbers of 0.95 and 1.2, however at a Mach number of 1.2, the elevator effectiveness appears to be about the same as the low speed value.

(f). The static stability at low lift coefficients increases as Mach number increase from 0.6 to 0.9 to twice the low speed value, and at a Mach number of 1.2, C_{m_α} appears to be three times the low speed value. Previous data has shown that the airplane becomes longitudinally unstable considerably below maximum lift coefficient, and that the lift coefficient for instability decreases as Mach number increases to 1.0. This boundary appears to follow the buffet boundary somewhat and is about 0.1 to 0.2 lift coefficient above the buffet boundary. At a Mach number of 1.25 a point of instability was reached at a lift coefficient of about 0.76.

(g). The buffet boundary for altitude of 40,000 feet decreases from a lift coefficient of 0.6 at a Mach number of 0.70 to about 0.44 at a Mach number of 0.85, then rapidly drops to a lift coefficient of about 0.15 at a Mach number of 0.98. The boundary rises rapidly to a value of 0.60 lift coefficient at a Mach number of 1.0. The buffet boundary appears to be constant at a lift coefficient of 0.60 in the supersonic range to 1.2, limit of tests at these lift coefficients. The severity of buffeting follows a similar pattern, being more intense at Mach numbers near 0.9 for a given lift coefficient. The buffeting tail loads at Mach numbers near 0.9 increase slightly to a lift coefficient of 0.4, and increased rapidly thereafter reaching buffeting incremental loads of 1500 pounds at a lift coefficient of 0.8 and 40,000 feet.

(h). The drag coefficients for the D-558-II airplane are somewhat less than for the X-1 airplane at comparable lift coefficients throughout the Mach number range except at high lift coefficients in the vicinity of the drag rise. The values of drag coefficient lack the "hump" that is characteristic of straight wing airplanes at Mach numbers near 1.0.

(i). Temperature measurements were made on the canopy glass, wing skin, and fuselage skin. The values of stagnation temperature and nose skin temperature agree fairly well with calculated values using method of T.N. 1725. The theoretical nose skin temperature was computed assuming a perfect cone. The maximum values of stagnation and nose skin temperature measured were 200 and 80 degrees, F, respectively at a Mach number of about 1.8 at 67,000 feet altitude.

3. The pilot is going to Wright Field next week for fitting of a pressure suit so that flights above 45,000 feet may be made.

Herman O. Ankenbruck
Aeronautical Engineer

HOA:mem

CC: NACA Headquarters (2)
Ames
Lewis
ChBuAer.
Navy Liaison, Muroc
Projects Engineer
Files

Edwards, California
November 21, 1951

MEMORANDUM for Chief of Research

Subject: Progress report for the D-558-II (144) research airplane for the period November 3 to November 16, 1951.

1. A flight was made on Tuesday, November 13. The maximum Mach number was 1.11 at 48,000 feet altitude. The flight was made primarily to obtain longitudinal stability and maneuvering loads data and aileron effectiveness in the Mach number range between 1.1 and 0.7.

2. Another flight, the 4th by NACA, was made on Friday, November 16, 1951. The maximum Mach number was about 1.65 at an altitude of about 60,000 feet. The flight was made primarily to obtain longitudinal stability, maneuvering loads, and dynamic lateral stability data at high Mach numbers.

3. The data obtained in these flights are being reduced and analyzed and preparations are being made for a 5th flight as soon as possible.

Herman O. Ankenbruck
Aeronautical Engineer

HOA:mem
DEB
WCW

cc: NACA Headquarters (2)
Ames
Lewis
ChBuAero
Navy Liaison, Edwards
Projects Engineer
Files

Document 9, Memo, Herman O. Ankenbruck, Aeronautical Engineer, To: Chief of Research, Subject: Progress report for the D-558-II (144) research airplane for the period November 3 to November 16, 1951, Date: November 21, 1951

Edwards, California
December 18, 1951

MEMORANDUM for Chief of Research

Subject: Progress report for the D-558-II (144) research airplane for the period December 1 to December 14, 1951

1. A preliminary flight test program is being carried out with the D-558-II (144) airplane to determine the limits of flight before beginning the formal program.

Status of Program

2. Data has been obtained that indicates that the dynamic lateral stability is very poor in the transonic and supersonic speed range; and at speeds above a Mach number of 1.4, that the static directional stability deteriorates rapidly as Mach number increases.

3. At present, ways are being devised for more satisfactory lateral characteristics. The most promising of these appears to be the installation of a rate autopilot to add damping in yaw. Methods of applying an automatic pilot (already constructed) to this task are being considered, and means of increasing the static stability are being sought. To aid in the analysis of the lateral stability, values of the moments-of-inertia about the Z and X axes will be measured.

Status of Airplane

4. No flights were made since November 16 due to a wet lake bed.

5. Work on a new rudder locking device is progressing.

Herman O. Ankenbruck
Aeronautical Engineer

HOA:mem

cc: NACA Headquarters (2)
Ames
Lewis
ChBuAero
Navy Liaison, Edwards
Projects Engineer
Files

Edwards, California
July 23, 1952

MEMORANDUM for Chief of Research

Subject: Progress report for the D-558-II (144) research airplane for the period June 28 to July 11, 1952.

1. A successful flight (number 8) was made on July 10, 1952 to an altitude of 55,000 feet and a maximum Mach number of 1.68. The flight was made primarily to obtain information on longitudinal stability and tail loads.

2. Flight number 9 is planned for the week of July 13. The purpose of the flight will be essentially the same as flight number 8.

3. The proposed research memorandum "Some Measurements of Aerodynamic Heating Obtained During Demonstration Flights of the Douglas D-558-II Airplane", by Ira P. Jones, is being modified according to comments. The proposed research memorandum "Some Measurements of the Buffet Region of a Swept-Wing Research Airplane During Flights to Supersonic Mach Numbers", by T. F. Baker, is being modified according to comments. The proposed report on Handling Qualities at Supersonic Speeds is being reviewed in the section.

Herman O. Ankenbruck
Aeronautical Engineer

HOA:mem

DBB

cc: NACA Headquarters (2)
Lewis
Ames
ChBuAero
Projects Engineer
Files

Edwards, California
December 11, 1952

MEMORANDUM for Chief of Research

Subject: Progress report for D-558-II research airplane (144) for the period November 1 to December 1, 1952.

1. Thus far during the year 1952, eight flights have been made, primarily to explore longitudinal stability and control, wing and tail loads, lift, drag, and buffeting characteristics at high lift and supersonic speeds. Some data were obtained on static directional stability and control, dynamic stability, rudder hinge moments, and pressures at the base and along the side rear of the fuselage at high subsonic and supersonic speeds. During maneuvers at supersonic speeds, reduction of static longitudinal stability followed by instability was encountered at moderate values of normal-force coefficient at low supersonic speeds, these values of C_{N_A} seemed to be somewhat higher than at high subsonic speeds, but appear to decrease with increasing supersonic Mach number.

2. The inboard fences have been removed from the wings of the airplane, and pressure orifices on the wing have been connected to recorders. The airplane is in stand-by condition, however recent rain has made the lake bed unusable for the present.

3. Status of Reports

Published:

"Some Measurements of Aerodynamic Heating Obtained During Demonstration Flights of the D-558-II Airplane", by Ira P. Jones.

"Some Measurements of the Flying Qualities of the Douglas D-558-II Research Airplane", by H. O. Ankenbruck and T. E. Dahlen.

Interlaboratory review completed:

"Some Measurements of the Buffet Region of a Swept-Wing Airplane During Flights to Supersonic Mach Number", by T. F. Baker.

"Maximum Altitude and Maximum Mach Number Obtained With the Modified Douglas D-558-II Research Airplane During Demonstration Flights", by T. E. Dahlen.

Reports in review:

"Determination of Longitudinal Stability In Supersonic Accelerated Maneuvers For the Douglas D-558-II Research Airplane", by H. O. Ankenbruck.

4. Data on buffeting, wing and tail loads, static directional stability, etc., are being reduced and analyzed.

5. It is anticipated that the next flights will be primarily to explore the directional stability and control and vertical tail loads in supersonic flight. The program of longitudinal maneuvers will be continued in order to obtain a complete breakdown of the stability and control characteristics along with lift, drag, buffeting, wing and horizontal tail load data. This program will include measurements of wing chordwise pressures at one spanwise station at supersonic speeds.

Herman O. Ankenbruck
Aeronautical Engineer

HOA:mh
JEB
WCW

cc: NACA Headquarters (2)
Lewis
Ames
ChBuAero
Navy Liaison, Edwards

Projects Engineer
Files

Edwards, California
October 6, 1954

MEMORANDUM for Research Airplane Projects Leader

Subject: Progress report for the D-558-II (144) research airplane for the period September 1 to September 30, 1954.

1. During this period two flights were made. One flight canceled because of a grounding order by Reaction Motors until the thrust bearing in the turbine pump could be inspected to insure that a double thrust bearing was installed.

Flight 44 was made September 17 to obtain pressure distributions and structural loads measurements at low supersonic and subsonic speeds.

Flight No. 45 was made September 22 to obtain power on dynamic stability data at subsonic speeds. This flight was made at 30,000 feet because of a leaking cabin pressure-seal that prevented going to the higher altitudes.

2. Reports in progress:

(a) The report entitled, "Determination of Longitudinal Handling Qualities of the D-558-II Research Airplane at Transonic and Supersonic Speeds", by H. O. Ankenbruck has been published as RM H54G29A.

(b) The report entitled, "Lateral Motions Encountered with the D-558-II All-Rocket Airplane During Exploratory Flights to a Mach Number of 2.0", by H. O. Ankenbruck and G. M. Wolowicz is at Langley awaiting publication.

(c) A prospective report entitled, "Wing Loads Measurements at Supersonic Speeds of the Douglas D-558-II Research Airplane", by G. H. Robinson, G. E. Cothren, and C. Pembo is being revised according to Editorial Committee comments for interlaboratory review.

(d) A general airspeed report to include maximum Mach number and altitude data obtained on the D-558-II airplane is being prepared for Station Editorial Committee.

(e) A prospective report entitled, "Flight Determined Pressure Distributions over a Section of the Wing of the D-558-II Research Airplane at Mach Numbers up to 2.0", by G. H. Jordan and E. R. Keener is being revised according to Editorial Committee comments for interlaboratory review.

Gareth H. Jordan
Aeronautical Research Scientist

GHJ:mh

cc: NACA Headquarters (4)
Lewis - Ames
Navy Liaison, Edwards
ChBuAero
Projects, Engineer
Files

B-1-2

Edwards, California
October 3, 1956

MEMORANDUM for Chief, High-Speed Flight Station

Subject: Progress report for the D-558-II (144) research airplane for the period September 1, to September 30, 1956.

    1. During this report period one flight was made with the airplane. Flight number 69 was made September 25, 1956, to obtain vertical tail loads and stability and control data at M=1.1. The data obtained during this flight are felt to be sufficient to complete the vertical tail loads program.

    2. Following this flight a routine turbine pump inspection is being accomplished prior to returning to flight status.

                            Gareth H. Jordan
                          Aeronautical Research Scientist

GHJ:jr

DEB

cc: NACA Headquarters (4)
    Lewis - Ames
    W. J. Underwood
    Navy Liaison, Edwards
    ChBuAero
    Projects, Engineer
    Files

Document 14, Memo, Gareth H. Jordan, Aeronautical Research Scientist, To: Chief, High-Speed Flight Station, Subject: Progress report for the D-558-II (144) research airplane for the period September 1 to September 30, 1956, Date: October 3, 1956

Edwards, California
December 4, 1956

MEMORANDUM for Chief, High-Speed Flight Station

Subject: Progress report for the D-558-II (144) research airplane for the period November 1, to November 30, 1956.

1. During this reporting period two research flights were made with the airplane. Flight numbers 72 and 73 were made November 1, and November 7, to obtain static and dynamic stability data at a Mach number of 1.50 at 60,000 feet altitude. These data are being reduced and analyzed.

2. Instrumentation has been installed to measure the overall noise level in the aft portion of the fuselage at supersonic speeds. This instrumentation will be given an operational check-out on the next flight and it is anticipated that one flight will be required to obtain the presently planned noise data.

3. Additional flights on the airplane were delayed because of the instrumentation installation and due to a shortage of liquid oxygen. Liquid oxygen is available again and it is anticipated that research flights will resume the first week in December.

Gareth H. Jordan
Aeronautical Research Engineer

GHJ:jr

JF
HMD

cc: NACA Headquarters (4)
Lewis - Ames
W. J. Underwood
Navy Liaison, Edwards,
ChBuAero
Projects, Engineer
Files

Edwards, California
January 8, 1957

MEMORANDUM for Chief, High-Speed Flight Station

Subject: Progress report for the D-558-II (144) research airplane for the period December 1 to December 31, 1956

1. During this report period two research flights were made with the airplane. Flights numbers 74 and 75 were made December 14 and December 20 to obtain dynamic stability data at $M = 1.4$ at 60,000 and 45,000 feet and to obtain overall sound pressure levels at subsonic and supersonic speeds at 45,000 feet. These data are being reduced and analyzed.

2. Research flights on this airplane are completed and research instrumentation is being removed from the airplane. A 30-minute inspection of the rocket engine is currently being made and the airplane will be returned to flight status to be used for approximately seven pilot-familiarization flights.

Gareth H. Jordan
Aeronautical Research Engineer

CHJ:pm

cc: NACA Headquarters (4)
Lewis - Ames
W. J. Underwood
Navy Liaison, Edwards
ChBuAero
Projects, Engineer
Files

Document 16, Memo, Gareth H. Jordan, Aeronautical Research Engineer, To: Chief, High-Speed Flight Station, Subject: Progress report for the D-558-II (144) research airplane for the period December 1 to December 31, 1956, Date: January 8, 1957.

Muroc, California
September 13, 1951.

MEMORANDUM for Chief of Research

Subject: Progress report for the D-558-II (37975) research airplane for the period August 25 to September 7, 1951.

1. The rough draft of a report on wing and section loads obtained by pressure distributions during the Douglas rocket-jet flights on the airplane up to a Mach number of 1.04 is being revised by the author after review in section.

2. Analysis of dynamic longitudinal stability data obtained during flight 4 is continuing. Data obtained during flights 7, 8, and 10 are being worked-up and analyzed.

3. September 4 and 5 the weight and balance and the pitching moment of inertia (I_y) were determined. Also on September 5 a strain gage check calibration was made on the horizontal tail.

4. No flights were made during this two week period. A flight scheduled for September 7 has been delayed due to a fuel leak on the P2B-1S.

James R. Peele
Aeronautical Research Scientist

JRP:mem

cc: NACA Headquarters (2)
Ames
Lewis
ChBuAero
Navy Liaison, Muroc
Projects Engineer
Files

Document 17, Memo, James R. Peele, Aeronautical Research Scientist, To: Chief of Research, Subject: Progress report for the D-558-II (37975) research airplane for the period August 25 to September 7, 1951, Date: September 13, 1951

Muroc, California
September 28, 1951

MEMORANDUM for Chief of Research

Subject: Progress report for the D-558-II (37975) research airplane for the period September 8 to September 22, 1951.

1. On September 18, the D-558-II was air-launched at about 33,000 feet. Due to a malfunction of the rocket system, the pilot was unable to fire the rockets and had to jettison the rocket fuel. The flight was continued on jet power at an altitude of 20,000 to 25,000 feet. An accelerated pitching maneuver (simulating a possible rapid pitch to high angle of attack during carrier launching) was attempted at an altitude of around 20,000 feet and a Mach number of about 0.38. The airplane with gear down, flaps half down and slats closed (inadvertently left locked) pitched to a high angle of attack and fell off into a spin. The pilot reported that the recovery from the spin was normal after the gear was raised. Instrumentation functioned properly and the data obtained are being prepared for work-up.

2. The rough draft of a report on wing and section loads obtained by pressure distributions during the Douglas rocket-jet flights of the airplane up to a Mach number of 1.04 is being revised by the author after review in section.

3. Wing-load-distribution, buffet-tail-load and wing-fuselage pitching-moment data and control information to a normal-force coefficient greater than 1.0 and a Mach number around 0.90 are being evaluated. The C_{N_A} greater than 1.0 was reached in a pitch-up after reaching longitudinal instability at a C_{N_A} of approximately 0.6. Previous data indicate that the longitudinal instability boundary varies from a C_{N_A} of approximately 1.0 at $M = 0.50$ to a C_{N_A} of about 0.6 at $M = 0.90$.

JRP:mom

James R. Peele
Aeronautical Research Scientist

cc: NACA Headquarters (2)
Ames
Lewis
ChBuAero
Navy Liaison, Muroc
Projects Engineer
Files

Edwards, California
November 6, 1951

MEMORANDUM for Chief of Research

Subject: Progress report for the D-558-II (NACA 145) research airplane for the period October 6 to October 19, 1951

1. During the flight performed on September 26, the instrument inverter on the airplane malfunctioned during the landing approach, causing a loss of the landing record. Subsequently, this inverter was replaced by one of greater capacity, and the instrumentation was checked out.

2. A rocket-jet flight was not scheduled before October 11 or 12, because the Edwards Air Force Base nitrogen evaporator was inoperative. A rocket-jet flight scheduled for October 11 or 12 was cancelled because of a shortage of peroxide and the priority being given to the D-558-II (NACA 144) airplane. As a result of this cancellation, outboard fences (at 0.73 b/2) similar to those tested in the Langley Stability Tunnel on the D-558-II at low speed were installed on the NACA 145 airplane to check the results of the tunnel tests and to evaluate the effects on the longitudinal instability experienced with the airplane at high Mach numbers. A jet flight was scheduled to determine the effects of the fences at speeds below $M = 0.8$.

3. On October 18, the phase 2 airplane was dropped from the P2B-1S launch airplane at 30,000 feet and proceeded to perform a number of accelerated turns with slats locked and unlocked at speeds from $M = 0.75$ down to $M = 0.5$. Also, 1g stalls were performed in the airplane clean and dirty conditions, and several maneuvers were performed to determine airplane lift-drag ratios at speeds slightly above landing speed. Processing of flight films showed that no records were taken during any of these maneuvers because of malfunction of instrumentation soon after take-off of the phase 2 airplane.

4. The pilot of the phase 2 airplane reported a marked improvement in the airplane longitudinal characteristics as a result of adding outboard fences. Only mild or no pitch-up was now discernible in the accelerated turns, although a somewhat controllable instability was noted. The poor lateral stability previously noted in 1g stalls was also much improved, although buffet appeared more severe as speed was decreased (possibly because the airplane could go to higher C_{N_A} and lower speeds more easily).

5. As a result of these promising characteristics reported by the pilot, a rocket-jet flight is planned for early next week, during which transonic longitudinal characteristics with the outboard fences will be evaluated and data should be obtained for the speed range covered in the preceding flight (with no data).

Jack Fischel
Jack Fischel
Aeronautical Stability and Control Scientist

JF:rer

cc: NACA Headquarters (2)
Ames
Lewis
ChBuAero
Navy Liaison, Muroc
Projects Engineer
Files

CONFIDENTIAL

Edwards, California
November 8, 1951

MEMORANDUM for Chief of Research

Subject: Progress report for the D-558-II (NACA 145) research
airplane for the period October 20 to November 2, 1951

1. A rocket-jet flight of the D-558-II airplane was scheduled for October 22 to determine the effects of outboard wing fences (at 0.73 b/2) on the longitudinal instability experienced with the airplane at high Mach numbers. The usual pre-flight, climb, and pre-drop procedure was completed for the D-558-II airplane as scheduled. Just prior to drop of the D-558-II airplane from the P2B-1S launch airplane, however, it was noted that the D-558-II hydraulic pressure was low and actuation of airplane components was sluggish. The flight was thereupon cancelled and the planes landed together (the D-558-II in captive position) on the Edwards Air Force Base runway.

2. Inspection of the D-558-II airplane revealed a hydraulic leak near the aileron snubber on the left wing. This leak was repaired, and a flight rescheduled for October 23. Unfavorably high winds and clouds caused cancellation of scheduled flights on October 23 and October 24, and heavy rains obviated the use of the lake for the remainder of the current period.

3. As a result of the adverse lake-bed conditions for flying, and the subsequent grounding of the airplane, instrumentation changes are being made to the research airplane. A 36-channel oscillograph is replacing the 18-channel oscillograph for recording wing stresses and tail loads and stresses, and other instrumentation is being checked.

4. Analysis of the data thus far obtained on the airplane shows that the variation of stick force per g of normal acceleration during accelerated turns increases rapidly from a value of about 15 lbs at $M = 0.8$ to about 40 lbs at $M = 0.9$ and about 120 lbs at $M = 0.98$. Also, the apparent stability parameter $\Delta\delta_e/\Delta C_{N_A}$ increases from a value of approximately $12°$ at $M = 0.8$ to approximately $19°$ at $M = 0.9$, and approximately $42°$ at $M = 0.98$. Elevator-impulse maneuvers produced values of C_{m_α}

CONFIDENTIAL

that were approximately the same as the values obtained in the wind tunnel for the same c.g. location. These values of C_{m_α} were negative over the entire Mach-number range and became more negative with increase of M.

Jack Fischel
Aeronautical Stability and Control Scientist

JF:rer

cc: NACA Headquarters (2)
Ames
Lewis
ChBuAero
Navy Liaison, Muroc
Projects Engineer
Files

CONFIDENTIAL
SECURITY INFORMATION

Edwards, California
December 18, 1951

MEMORANDUM for Chief of Research

Subject: Progress report for the D-558-II (NACA 145) research airplane for the period December 1 to 14, 1951.

1. A rocket-engine run-up, following the rocket-engine inspection completed in the preceding reporting period, was successfully completed and preparations made for a flight on December 4 or 5. However, heavy rains, which obviated use of the lake bed, caused cancellation of this flight. Plans were subsequently made for a flight during the week of December 10, but additional heavy rains during this week caused further postponement of flight tests.

2. Because of the grounding of the airplane, preparations are continuing for installation of tufts on the right wing panel, and cameras are being installed in the vertical fin and forward fuselage section of the airplane. In addition, the instrumentation section is building a total-head and angle-of-attack measuring mechanism to be mounted on a boom immediately ahead of the horizontal tail, and which will provide dynamic pressure and downwash measurements in this vicinity.

3. Further analysis of data obtained in preceding flights is continuing. Analysis of wing pressure measurements, tail load data, and stability and control data indicate that in a high-Mach-number accelerated turn during which a pitch-up was experienced, a sizeable decrease in wing stability, and particularly in wing-fuselage stability, was noted at the point where pitch-up occurred. This is thought to result from an unloading of the tip sections of the wing as C_{N_A} increased, and from an unstable trend in wing-fuselage pitching moment. Also, at this high Mach number, it is thought that an increase in downwash at the tail had a secondary effect in producing the pitch-up.

In low-Mach-number accelerated turns, the wing-fuselage, which is unstable at low values of C_{N_A}, becomes stable at high values of C_{N_A}. In this speed range, the pitch-up experienced is thought to result almost entirely from an increase in downwash angle as a increased and the tail entered the wing wake. Therefore, an effort will be made, as described in a preceding paragraph, to obtain downwash measurements at the tail of the airplane and to determine these effects on the stability of the airplane.

Jack Fischel
Aeronautical Stability and Control Scientist

JF:mem
WCW

cc: NACA Headquarters (2)
Ames
Lewis
ChBuAero
Navy Liaison, Edwards
Projects Engineer
Files

CONFIDENTIAL
SECURITY INFORMATION

Edwards, California
July 23, 1952

MEMORANDUM for Chief of Research

Subject: Progress report for the D-558-II (NACA 145) research airplane for the period June 28 to July 11, 1952.

1. A rocket-thrust calibration was performed on the Edwards thrust stand, and preparations were made for a rocket-jet flight to obtain slat load data and stability data with the slat in the locked-open position. This flight was to extend the Mach number range of data obtained with the same configuration in the preceding flight with jet-engine power alone.

2. On July 3, 1952 a rocket-jet flight was accomplished with the foregoing configuration at Mach numbers up to approximately 0.96. Maneuvers completed consisted of accelerated turns and 1g stalls, and the pilot reported no instability, only changes in stability, up to high accelerations. Because of a failure in the inverter, however, all internal instrumentation failed to record, with the exception of the oscillograph and manometer records. Slat loads data are being worked-up with the aid of cockpit photographed instrumentation.

3. The airplane was subsequently prepared for captive flight to Ames Laboratory for display at the NACA Annual Inspection, was flown there on July 9, and should return prior to July 18.

4. Structural design of the chord extensions is continuing.

5. Work-up is continuing on tail-load, wing-pressure, and stability and control data obtained during the preceding slat-open flight and during preceding flights in the original airplane configuration.

Jack Fischel
Jack Fischel
Aeronautial Stability and Control Scientist

JF:mem
DEB

cc: NACA Headquarters (2)
Lewis
Ames
ChBuAero
Navy Liaison, Edwards
Projects Engineer - Files

Edwards, California
August 15, 1952

MEMORANDUM for Chief of Research

Subject: Progress report for the D-558-II (145) research airplane for the period July 26 to August 8, 1952.

1. A jet-rocket flight of the airplane with the slats locked-open was scheduled for July 28, but was postponed to July 31 because of inclement weather. On July 31, a jet-rocket flight (flight 17) was performed during which a 2-rocket climb was performed after drop from the P2B-1S airplane; however, because of faulty canopy heating, the canopy of the D-558-II airplane iced-over at about 35,000 feet altitude and the rocket climb and acceleration was terminated. The rocket engines were subsequently shut-off, rocket fuels were jettisoned, and altitude was lost to deice the canopy. Below 25,000 feet altitude, the canopy cleared and several maneuvers were performed. Data obtained on this flight are being worked-up.

2. Preparations were made again to obtain a jet-rocket flight with the slats locked-open, but because of engine trouble on the P2B-1S launch airplane, this flight (flight 18) did not take place until August 8. A 2-rocket climb to 35,000 feet, followed by an acceleration to $M \approx 0.96$, was performed, and subsequent maneuvers consisted of accelerated turns at speeds up to $M \approx 0.96$, as well as several rolls and sideslips at $M \approx 0.7$. Data film from this flight are being developed.

3. Work-up is continuing on stability and control, wing pressure, and tail load data obtained during preceding flights. Preparation of a report on changes in longitudinal stability encountered at maneuvering lift coefficients up to high subsonic speeds is also continuing.

4. Preliminary analysis of the data obtained with slats locked-open indicates improved stability characteristics for the airplane as compared to the original slats locked-closed configuration — particularly as regards the severity of the pitch-up encountered at high values of lift coefficient. In addition, the airplane with slats locked-open appears to have higher values of C_D at low values of C_L, and lower values of C_D at high values of C_L, than the airplane with slats closed.

Jack Fischel
Jack Fischel
Aeronautical Stability and Control Scientist

JF:mem
DEB
WCW

cc: NACA Headquarters (2)
Lewis
Ames
ChBuAero
Navy Liaison, Edwards
Projects Engineer
Files

CONFIDENTIAL

Total 2 pages

B-1-2
D-558
A-7

Edwards, Calif.
November 21, 1952

MEMORANDUM for Chief of Research

Subject: Progress report for the D-558-II (145) research airplane for the period October 1 to November 1, 1952.

1. Tests of the airplane with wing slats locked in the half-open position and with no stall-control fences on the wing were made on October 8. This was flight 20, a rocket-jet flight, and accelerated longitudinal maneuvers were performed up to $M \approx 0.97$ to determine any improvement in the longitudinal stability characteristics of the airplane as a result of locking the slats half-open. Because a relay in the instrumentation circuit became jammed, only the first two high-speed maneuvers were recorded. However, the pilot reported pitch-ups were encountered during the accelerated maneuvers, and the recorded data verified this. At the highest speed tested, the plane pitched to a value of $\alpha = 36°$ and $C_{N_A} = 1.79$.

2. The airplane was subsequently modified by placing the slats in their former position (locked or free-floating) in order to obtain data on the airplane in the basic configuration (clean condition, no wing fences).

3. On October 22, a rocket-jet flight (flight 21) was performed with the airplane in the basic configuration. Maneuvers performed included accelerated longitudinal maneuvers (from $M \approx 0.45$ to $M \approx 0.97$), 1g stalls, and aileron rolls. After the first two high-speed turns, the elevator position was not recorded, due to breakage of the string connecting the elevator and the control-position-transmitter. The pilot reported pitch-ups during the turns, especially at high speed, where lateral and directional oscillations were simultaneously encountered after the start of the pitch-up.

4. Because of rough operation of the jet engine during the last flight, the airplane was grounded for removal and replacement of the jet engine. Simultaneously, the rocket engine was removed for inspection, and chord extensions are being installed on the outer 0.32 b/2 of the wings. It is also planned to install the drogue chute attachments on the airplane at this time, as well as angle-of-attack and total-head measuring equipment on the horizontal tail. The instrumentation group is meanwhile replacing the 60-cell manometers with new 24-cell manometers, and performing other work required on the airplane.

5. Work-up is continuing on load data, lift and drag data, stability and control data, and wing-pressure data obtained during preceding flights.

6. The rough draft of a report on measurements of longitudinal stability at transonic speeds, including the effects of outboard wing fences, is currently being prepared for transmittal to other laboratories for review.

CONFIDENTIAL
SECURITY INFORMATION

The rough draft of a report on buffeting characteristics of the airplane up to M = 0.95, and of a report on wing-span loading characteristics through a range of C_N and at 3 speeds (subsonic to transonic), obtained from wing-pressure measurements, is currently being reviewed in section.

Jack Fischel
Aeronautical Stability and Control Scientist

JF:mem

cc: NACA Headquarters (2)
    Lewis
    Ames
    ChBuAero
    Navy Liaison, Edwards
    Projects Engineer
    Files

Total 3 pages
REC'D DEC 15 1952

Edwards, California
December 11, 1952

MEMORANDUM for Chief of Research

Subject: Progress report for the D-558-II (145) research
airplane for the period November 1 to December 1, 1952.

1. The airplane is currently having pressure instrumented wing chord extensions (shown by wind-tunnel tests to improve the longitudinal stability characteristics) installed over the outer 0.32 wing semispans. In addition, drogue chute attachments are being installed at the tail of the airplane in order to make feasible landings on the runway while the lakebed is wet and unusable. A new jet engine has been installed in the airplane, and the rocket engine is to be reinstalled shortly.

2. The instrumentation group is installing new 24-cell manometers and replacing the existing angle-of-attack and sideslip angle systems.

3. During the past several months, the airplane has been flown in various wing slat and wing fence configurations in attempts to improve the longitudinal stability characteristics of the airplane and eliminate the pitch-up encountered with the airplane in the original configuration (slats closed and inboard fences on wings) as well as to obtain loads on the fully-extended slat up to large values of C_{N_A} and Mach number. The configurations flown are as follows:

(a) Slats fully-extended and inboard fences on wing.
(b) Slats fully-extended and inboard fences removed from wing.
(c) Slats half-extended and no wing fences.
(d) Slats closed and no wing fences.

Data were obtained during the course of these flights up to $M \approx 0.97$ in accelerated longitudinal maneuvers, aileron rolls, and gradual sideslips. Almost all the stability and control and tail-load data have been worked-up, and all the strain-gage slat-load data have been worked-up; however, only some of the slat-pressure data have been worked-up. All lift-drag data have been worked-up.

4. The data showed that the configurations flown with slats fully-extended were an improvement over the original airplane configuration, inasmuch as the pitch-up previously encountered appeared to be eliminated or alleviated, except at $M \approx 0.8$ to 0.85. Stability changes were still encountered over

CONFIDENTIAL

the C_{N_A} or α range, but the most apparent change to the pilot was the stick-free instability encountered at moderate values of C_{N_A} or α. In either configuration (c) or (d) above, the airplane appeared to be as longitudinally uncontrollable as in the original configuration, after the pitch-up occurred. Drag with slats fully-extended was higher at low values of C_L and lower at large values of C_L than in the original configuration. Slat loads, with slat fully-extended, tended to increase more rapidly than C_{N_A} as α increased, and the variation of slat normal-force coefficient with airplane C_{N_A} tended to decrease as M increased.

5. The status of several prospective reports based on results obtained on the original configuration and with added outboard fences as well as the configurations listed in 3(a) to (d) above, are as follows:

(a) "Transonic Flight Determination of the Longitudinal Stability in Accelerated Maneuvers for the Douglas D-558-II Research Airplane Including the Effects of an Outboard Wing Fence" by Jack Fischel and Jack Nugent - being reviewed at other NACA laboratories.

(b) "Longitudinal Trim and Control, and Lateral Trim Characteristics at Transonic Speeds of Douglas D-558-II Research Airplane, Including Effects of Outboard Wing Fences", by Jack Fischel - Rough draft not yet started.

(c) "Some Measurements of Buffeting Encountered by a Douglas D-558-II Research Airplane in the Mach Number Range from 0.5 to 0.95", by Thomas F. Baker - being reviewed in section.

(d) "Variation with Lift of the Relative Span Loading of the Douglas D-558-II Research Airplane for Mach Numbers of 0.59, 0.76, and 0.89", by Sol Tenenbaum - being reviewed in section.

(e) "Pressure Measurements over the Wing of the Douglas D-558-II Research Airplane at Level Flight Lifts to Mach Number of 1.14", by James R. Peele - being revised by author after review in section.

(f) "Transonic Flight Measurements of the Aerodynamic Loads on the Extended Slat of the Douglas D-558-II Research Airplane", by James R. Peele.- Rough draft almost ready for section review.

(g) "Lift and Drag Results of the Douglas D-558-II Research Airplane with the Slats Fully Extended and Retracted", by Jack Nugent - Rough draft being reviewed in section.

(h) "Effects of Wing Slats and Inboard Wing Fences on the Transonic Longitudinal Stability Characteristics of the Douglas D-558-II Research Airplane in Accelerated Maneuvers", by Jack Fischel - Rough draft being prepared.

Jack Fischel
Aeronautical Stability and Control Scientist

JF:mh

DEB

WCW

cc: NACA Headquarters (2)
Lewis
Ames
ChBuAero
Navy Liaison, Edwards

Projects Engineer
Files

Edwards, California
December 20, 1954

MEMORANDUM for Research Airplane Projects Leader

Subject: Progress report for the D-558-II (145) research airplane for the period November 1 to November 30, 1954.

1. No flights were obtained during this period with the stores or pylon configurations because the large stores-loads measurements reported in the preceding period were being investigated. While additional data work-up for the stores flights was being performed at HSFS, Douglas, El Segundo engineering personnel were contacted to obtain more detailed information on the limit-load characteristics of the wing-pylon-large stores configuration. Detailed information on stores-loads measurements obtained in a wind-tunnel investigation by Douglas would also indicate the magnitude of loads that might be measured on the D-558-II airplane in flight, and indicate the feasibility of utilizing the present stores configuration. Douglas personnel are gathering this information and will relay it to HSFS personnel soon.

2. If further flights of the present stores configuration are deemed feasible, a complete calibration of the pylon-mounted strain gages measuring the pylon-stores loads will be performed prior to resumption of the flight program.

3. A preliminary evaluation of the buffeting characteristics of the airplane with the pylon and large stores (from the c.g. accelerometer) indicates the C_{N_A} - M boundary for the onset of buffeting is approximately the same as for the clean airplane at Mach numbers up to about 0.95. Apparently the increase in buffeting felt by the pilot in accelerated longitudinal maneuvers with the stores may have been the stores shaking.

4. Work-up of data obtained in previous flights is continuing.

Jack Fischel
Aeronautical Research Scientist

JF:mh

cc: NACA Headquarters (4)
Lewis - Ames
Navy Liaison, Edwards
ChBuAero
Projects, Engineer
Files

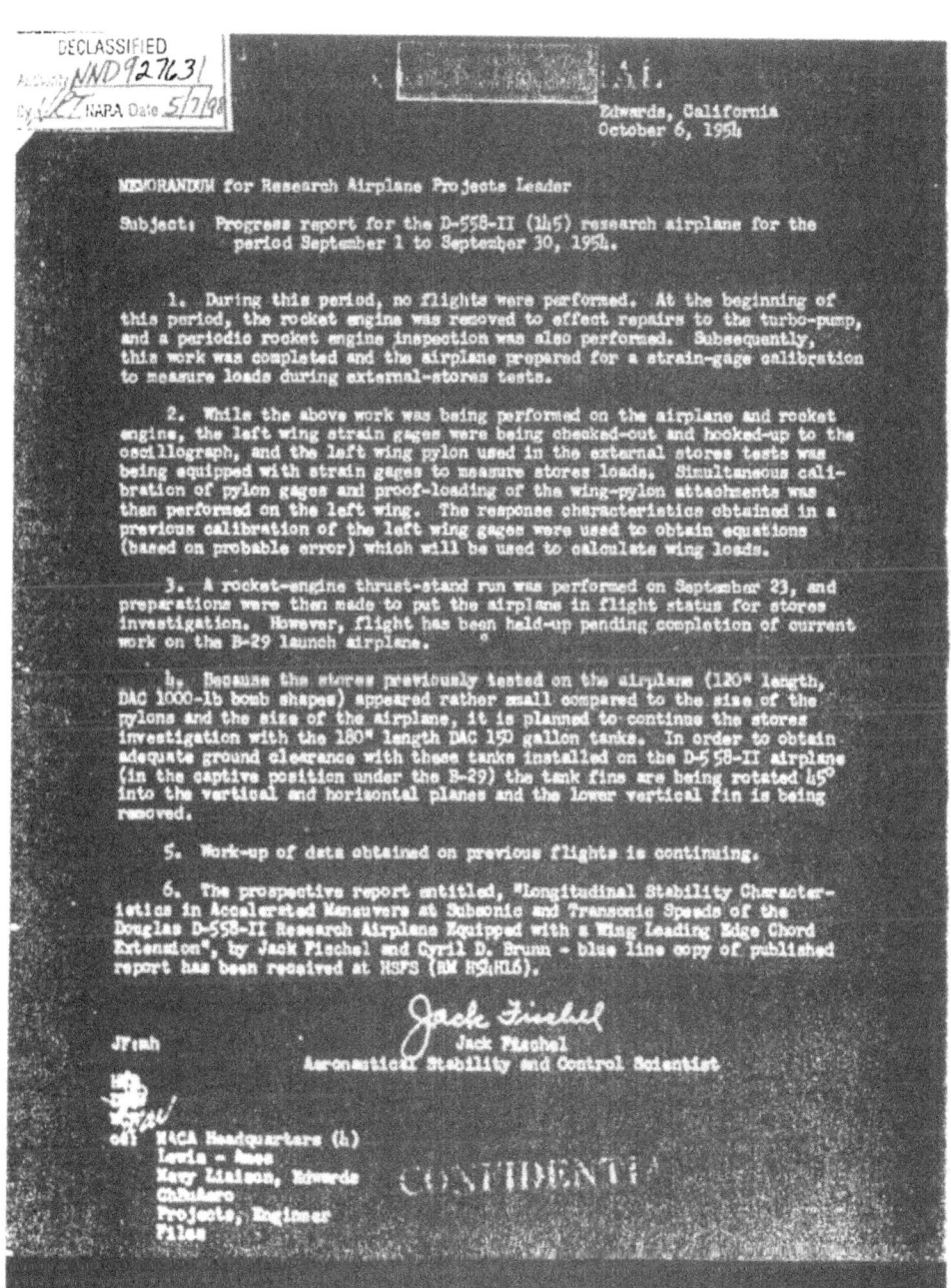

Document 27, Memo, Jack Fischel, Aeronautical Stability and Control Scientist, To: Research Airplane Projects Leader, Subject: Progress report for the D-558-II (145) research airplane for the period September 1 to September 30, 1954, Date: October 6, 1954

Edwards, California
April 18, 1955

MEMORANDUM for Research Airplane Projects Leader

Subject: Progress report for the D-558-II (145) research airplane for the period March 1 to March 31, 1955.

1. No flights were obtained during this period. Initially, operational and instrumentation work were performed on the airplane to prepare the airplane for flight. Among the operational work performed was ordinary maintenance and re-installation of the overhauled jet and rocket engines. However, during pre-flight ground checks, it was determined that both engines were malfunctioning; and subsequent examination revealed a leak at the lox valves of the rocket engine and a faulty starter on the jet engine. Both engines were then removed for rework, and are currently being re-installed in the airplane. During this period, calibration of all NACA instrumentation was completed. Plans are being made for resumption of the external-stores flight program within the next week.

2. Gage equations, for the left-pylon strain gages measuring store-pylon loads, were worked up from the calibration performed and are being used to reduce the loads data obtained during 2 previous flights with the DAC 180" length 150-gallon stores.

3. A prospective report entitled, "Effect of Several Wing Modifications on the Low-Speed Stalling Characteristics of the Douglas D-558-II Airplane", by Jack Fischel and Donald Reisert, was transmitted to other NACA laboratories for review.

Jack Fischel
Aeronautical Research Scientist

JF:mh

cc: NACA Headquarters (4)
Lewis - Ames
Navy Liaison, Edwards
ChRuAero
Projects, Engineer
Files

Edwards, California
August 11, 1955

MEMORANDUM for Research Airplane Projects Leader

Subject: Progress report for the D-558-II (145) research airplane for the period July 1 to July 31, 1955.

    1. No flights were performed with this airplane during this period because the airplane has been grounded for operational work. The rocket engine and pump were removed for periodic inspection and re-installed in the airplane. Periodic inspection of the stabilizer and stabilizer operating mechanism has been delayed while a stabilizer loading jig is fabricated for use during the inspection procedure. This inspection should be completed within the next few days, to be followed by resumption of the external-stores flight test program.

    2. Handling qualities data, buffeting data, wing and pylon-stores loads data, and lift and drag data previously obtained during external-stores flight tests are currently being reduced.

                                            Jack Fischel
                                       Aeronautical Research Scientist

JF:ta

cc: NACA Headquarters (4)
    Lewis - Ames
    Projects, Wright Patterson
    W. J. Underwood
    Major William W. Penn, Jr.
    Projects, Engineer
    Files

CONFIDENTIAL

Document 29, Memo, Jack Fischel, Aeronautical Stability and Control Scientist, To: Research Airplane Projects Leader, Subject: Progress report for the D-558-II (145) research airplane for the period July 1 to July 31, 1955, Date: August 11, 1955

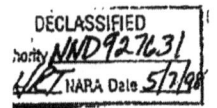

Edwards, California
July 24, 1956

MEMORANDUM for Chief, High-Speed Flight Station

Subject: Progress report for the D-558-II (145) research airplane for the period June 1, to June 30, 1956.

    1. Repairs to the B-29 launch airplane, required, as a result of damage sustained during March, were virtually completed during this period. Prior to checkout of the B-29 airplane in preparation for completion of the one to two flights remaining of the D-558-II test program, it was noted that the number 1 engine of the B-29 appeared defective, so an engine change will be performed. Hence, no flights were performed with the D-558-II (145) airplane during the current report period.

    2. Analysis of stability, performance and loads data obtained during preceding flights with external stores is continuing as other work permits. These data are being included in reports being prepared.

                                        Jack Fischel
                            Aeronautical Research Scientist

JF:jr

HMD

DEB

WCW

cc: NACA Headquarters
    Lewis - Ames
    W. J. Underwood
    Navy Liaison, Edwards
    ChBuAero
    Projects, Engineer
    Files

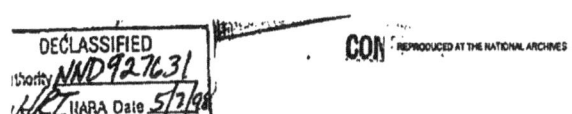

August 13, 1951

Major General Frederick R. Dent, Jr., USAF
Commanding General
Wright Air Development Center, WCC
Wright-Patterson Air Force Base
Dayton, Ohio

Dear General Dent:

As part of the Air Force-Navy-NACA joint research airplane program, the NACA will soon receive the Navy D-558-II all-rocket airplane and the X-2 and X-3 airplanes. Since much of the flight testing on these airplanes will be done at extremely high altitudes, it is believed mandatory, from the standpoint of pilot safety, that the pilot be equipped with a pressure suit.

The pressure suit presently under research and development by the Air Force appears ideal for this purpose. It is requested, therefore, that the Air Force loan or assign to the NACA, a pressure suit and the attendant auxiliary equipment.

Mr. A. Scott Crossfield, the NACA test pilot scheduled to fly the above noted airplanes, will be at the Wright Air Development Center on August 16, or 17, 1951. It would be greatly appreciated if Mr. Crossfield could be fitted for the pressure suit at this time.

Sincerely yours,

Ira H. Abbott

Ira H. Abbott
Assistant Director
for Research

LES:bfm

Copy to Langley

Copy to Mr. Underwood

CONFIDENTIAL

Document 31, Letter, Ira H. Abbott, [NACA] Assistant Director for Research, To: Major General Frederick R. Dent, Jr., USAF, Commanding General, Wright Air Development Center, August 13, 1951

October 23, 1951

From: National Advisory Committee for Aeronautics

To: Chief, Bureau of Aeronautics, Department of the Navy, Washington 25, D. C.

Attn: Ae-5

Subject: Request for assignment of Navy omnienvironment full-pressure pilot suit to NACA pilot of D-558-II airplane

Reference: BuAer letter, July 9, 1951, Aer-DE-3 08791

1. Full exploitation of the capability of the D-558-II airplane in the NACA's research program endorsed by the Bureau of Aeronautics in the reference letter will require operation at altitudes so high that the available emergency equipment is not adequate to protect the pilot in the event of loss of pressurization. Because of the seriousness of this problem, the NACA has investigated known developments in pressurized suits. It has been concluded that the omnienvironment full-pressure suit developed by the Navy Aero Medical Equipment Laboratory, Project No. TED NAM-GB-519057, offers the most suitable solution for this problem for some time to come.

2. NACA representatives have discussed the problem of protecting the D-558-II pilot, as well as the state of development of the Navy pressurized suit, with personnel of the Bureau of Aeronautics and NAMEL. These working level discussions have indicated that immediate assignment of a Navy pressurized suit to the NACA's D-558-II pilot will provide the needed protection and at the same time provide NAMEL with a useful field test of the equipment. It has been determined that the D-558-II cabin pressurization system is very adaptable to this suit.

3. The NACA is aware that the omnienvironment suit is still in its development stage and expects the Bureau of Aeronautics to accept no responsibility in the event of malfunction or failure of the suit. The protection which the suit will provide the Committee's pilot, however,

CONFIDENTIAL

Aeronautics — October 23, 1951

is so needed that the NACA urgently requests that the Bureau of Aeronautics arrange to provide NACA pilot Scott Crossfield with an omnienvironment full-pressure suit at the earliest possible date. The NACA would appreciate receiving the Bureau's views concerning this request.

J. W. Crowley
Associate Director
for Research

CLoW:bbn

Copy to Langley - Attn: Mr. H. A. Soulé

Copy to Muroc (by hand by Mr. S. Crossfield)

CONFIDENTIAL

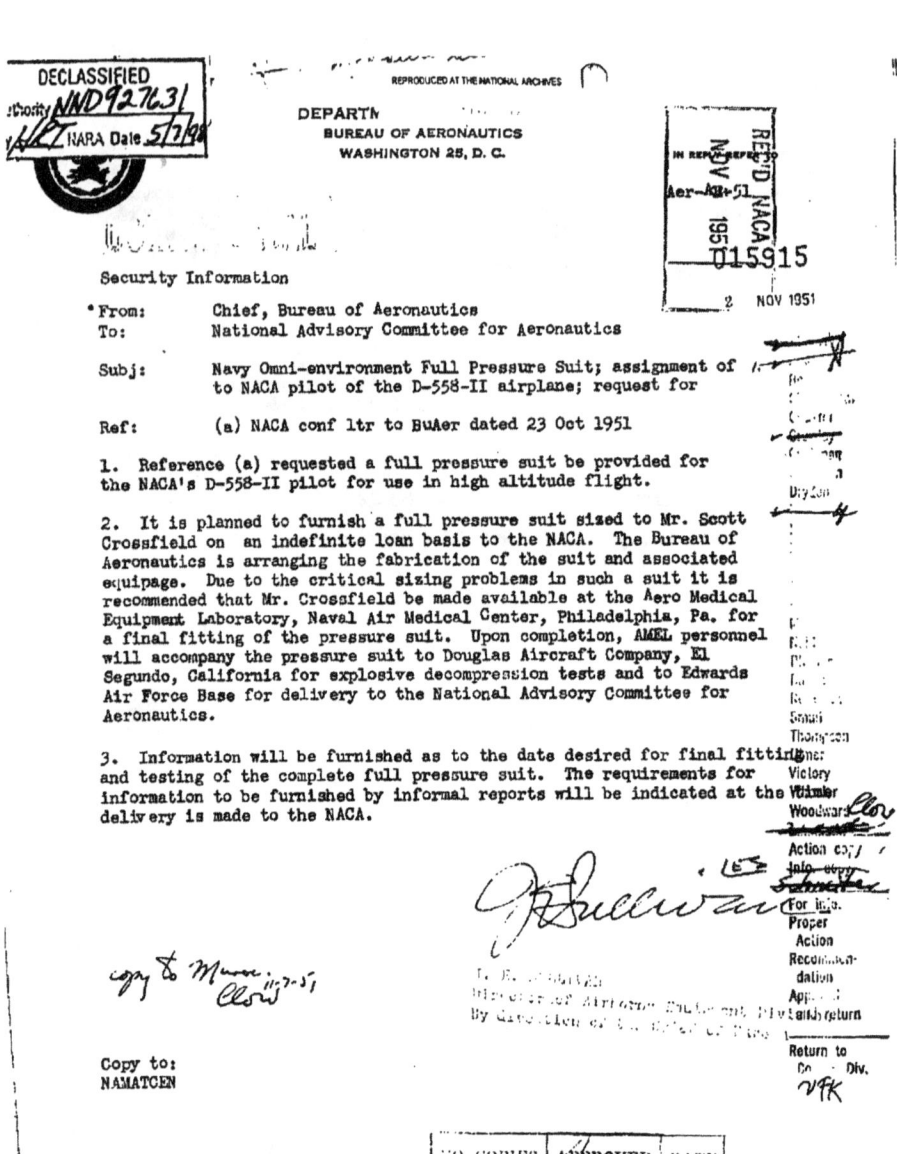

DEPARTM
BUREAU OF AERONAUTICS
WASHINGTON 25, D. C.

Security Information

From: Chief, Bureau of Aeronautics
To: National Advisory Committee for Aeronautics

Subj: Navy Omni-environment Full Pressure Suit; assignment of to NACA pilot of the D-558-II airplane; request for

Ref: (a) NACA conf ltr to BuAer dated 23 Oct 1951

1. Reference (a) requested a full pressure suit be provided for the NACA's D-558-II pilot for use in high altitude flight.

2. It is planned to furnish a full pressure suit sized to Mr. Scott Crossfield on an indefinite loan basis to the NACA. The Bureau of Aeronautics is arranging the fabrication of the suit and associated equipage. Due to the critical sizing problems in such a suit it is recommended that Mr. Crossfield be made available at the Aero Medical Equipment Laboratory, Naval Air Medical Center, Philadelphia, Pa. for a final fitting of the pressure suit. Upon completion, AMEL personnel will accompany the pressure suit to Douglas Aircraft Company, El Segundo, California for explosive decompression tests and to Edwards Air Force Base for delivery to the National Advisory Committee for Aeronautics.

3. Information will be furnished as to the date desired for final fitting and testing of the complete full pressure suit. The requirements for information to be furnished by informal reports will be indicated at the delivery is made to the NACA.

Copy to:
NAMATCEN

Document 33, Letter, J.E. Sullivan, Director of Airborne Equipment Division, [Navy] Bureau of Aeronautics, To: National Advisory Committee for Aeronautics, Subject: Navy Omni-environment Full-pressure Suit, assignment of to the NACA pilot of the D-558-II airplane, November, 1951

NAVAL AIR MATERIAL CENTER
Naval Air Experimental Station
Philadelphia 12, Pennsylvania

XG-2-WBC:beb
J26-2
01024

8 DEC 1953

CONFIDENTIAL

From: Director, Naval Air Experimental Station
To: Chief, Bureau of Aeronautics (AE-513)

Subj: TED No. NAM AE 5101 Omni-environment full pressure suit, research, development and test of; Flight test in D-558-II airplane at Edwards Air Force Base, Edwards, California 3 August to 4 September 1953; report on

Ref: (a) BUAER conf ltr Aer-AE-513, 011176 of 9 Jul 1953

Encl: (1) NAXSTA photographs 280279, 280280, 280281, 280282, 280283, 280284 and 280285 showing the David Clark Co. full pressure suit and back pan type gas regulating equipment.

1. In compliance with reference (a), an Aeronautical Medical Equipment Laboratory representative was ordered to Edwards Air Force Base to serve as technical advisor to Lt. Col. M. E. Carl, USMC, and the NACA on the operation and installation of the David Clark type full pressure suit and Firewel Co. control equipment in the D-558-II rocket airplane.

2. Lt. Col. Carl made six (6) attempts to fly the D-558 airplane in the full pressure suit with a back pan control system constructed at the AMEL. The suit and back pan gas regulating equipment are shown in enclosure (1). The D-558-II was launched from the bottom of a B-29 "mother" ship at altitudes of approximately 33,000 feet. The dates for these attempts were 13, 14, 18, 21, 31 August and 2 September 1953. All, except the August 13 launch, were successful. The August 13 Launch was cancelled prior to drop due to the loss of the D-558-II oxygen supply through a "cocked" oxygen disconnect. This malfunction was corrected and did not reoccur on any of the subsequent flights.

3. On the 2 August flight, Lt. Col. Carl flew to 83,235 feet for a new world altitude record and on the 2 September flight established a new military speed record of 1143 mph.

4. Mr. J. Ruseckas of David Clark Co., Mr. Scott Crossfield of NACA, and Mr. W. B. Cassidy of the AMEL accompanied Lt. Col. Carl in the B-29 to assist in the donning of the suit and connecting the pressurization and respiration services in the D-558-II. This was accomplished between

CONFIDENTIAL

Encl (2)

XG-2-WBC:beb
J26-2
01024

approximately 1,000 feet and 10,000 feet. A ground crew met Lt. Col. Carl at the completion of each flight to disconnect ship service lines to the suit and assist in the removal of the pressure suit and personal gear.

5. The flights were made in the unpressurized condition and at no time during any of the flights was the suit pressurized above the pressure needed for ventilation. This pressure was approximately 0.5 psi. Compressed nitrogen was used as the ventilation medium in all flights. Lt. Col. Carl controlled the amount of ventilation by means of a hand operated needle valve. On two (2) of the five (5) successful flights, the pressure suit was dry when Lt. Col. Carl returned to the ground; on one (1) it was damp; and on the other two (2) it was very wet. The factors which appeared to effect Lt. Col. Carl's comfort and the amount of sweating were as follows:

    a. Outside temperature - Early morning flights were much more desirable than any other time of the day because of the favorable temperatures outside and inside the B-29.

    b. Amount of time spent in the suit without ventilation. This time was kept to a minimum by fast preparation and proper flight scheduling.

    c. Emotional status of pilot.

    d. Amount of physical exertion by the pilot while wearing the suit.

    e. Distribution, quantity and temperature of ventilating nitrogen.

6. The following comments regarding the suit and component equipment reflect the opinions of Lt. Col. Carl, and the NACA personnel closely associated with these flight tests and the AMEL and David Clark Co. representatives:

    a. The nylon gloves were not rubberized sufficiently to provide the grippage necessary to efficiently operate toggle switches, buttons, and levers.

    b. The oxygen valving for the defogging mechanism on the suit requires refining to provide a "fool proof" system.

    c. Since there was no ventilation garment with the suit, the distribution of the ventilation air was erratic and not dependable.

    d. The visibility in all directions other than straight ahead was poor. The inability to look down (front, left and right) especially annoyed the pilot because it limited his view of the instrument panel and control mechanism.

CONFIDENTIAL

XQ-2-WBC:bcb
J26-2
01024

e. The suit was tailored too tightly. A looser fitting suit would permit easier donning with little sacrifice of mobility. This would also reduce wear.

f. The accessory equipment, i.e., regulators, controllers, bailout supply cylinders, parachute, etc., should be reduced in size and weight and better integrated.

g. There should be a flight indoctrination in the TV-2 or a similar trainer. This is considered essential for all pilots before they are permitted to fly single place aircraft wearing the full pressure suit.

h. The pressure suit should be washed, dried and powdered after each wearing.

i. A device should be designed to accompany each suit or group of suits in the field for measuring suit leakage rate.

j. A portable test stand should be designed for conducting pre-flight checks of the suit controller, breathing regulator, bailout cylinder, pressure reducer and flow check valves.

k. A need for a more suitable disconnect to connect pressure suit and component equipment in the airplane is apparent.

F. A. SANTNER
By direction

CONFIDENTIAL

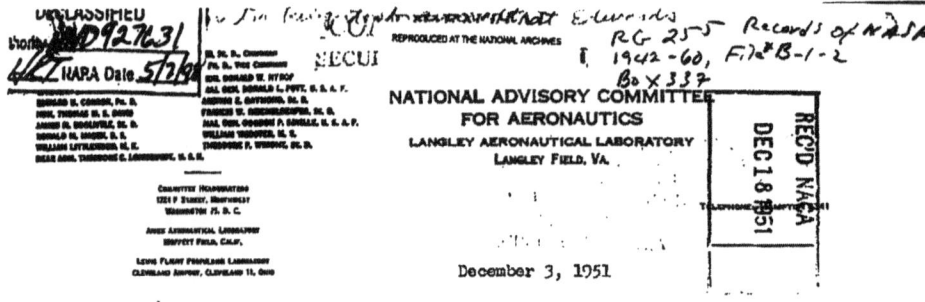

NATIONAL ADVISORY COMMITTEE
FOR AERONAUTICS
LANGLEY AERONAUTICAL LABORATORY
LANGLEY FIELD, VA.

December 3, 1951

From: NACA HSFRS-Edwards
To: NACA

Subject: Choice of color for Research Aircraft at Edwards

Reference: NACA letter to HSFRS dated November 15, 1951: Clow. mc.

The reference letter transmitted an inquiry from the Department of the Navy, Bureau of Aeronautics, regarding the optimum color for NACA research aircraft undergoing tracking tests at Edwards.

To realize the accuracy desirable from the radar tracking data, it is important that, in addition to being readily visible for optical tracking purposes, the airplane be photographable over the entire range of its flight path where data is desired. One of the early aircraft to undergo tracking tests at Edwards was the D558-1; this airplane was originally painted red. It was found during these initial tests that very little photographic contrast was being obtained between the red airplane and the relatively dark blue sky conditions prevalent in this area with the result that photographs could not be obtained to ranges greater than the order of 25 to 30 thousand yards. It was reasoned that the photographic contrast could be increased by using the lightest color possible against the darker sky. On this basis, the aircraft was test painted white and both visibility and photographability were found to be greatly increased. With proper filtering techniques, photographs of the airplanes are now taken to greater than 60,000 yards and are generally visible over their entire test flight range.

Fluorescent paints were briefly investigated to ascertain their value for our specific requirements but were not found to excel white from a visual stand point while photographic range was decreased to a value slightly greater than the original D558-1 color.

It should be pointed out that while white has proved superior photographically and visually for the type of tests and sky conditions here at

CONFIDENTIAL
SECURITY INFORMATION

Edwards, the same may not apply where lighter sky backgrounds due to haze conditions are existant.

Walter C. Williams
Chief, High-Speed Flight Research Station

GMT:dlh

Carbon copy to Mr. Soulé

CONFIDENTIAL
SECURITY INFORMATION

Memorandum for Gordon S. Williams
                 News Bureau Manager
                 Boeing Airplane Company
                 Seattle, Washington

Use of the air-launch technique in connection with flights of high-speed research airplanes should not be considered in the same light as the parasite fighter which is carried to a theater of operations, released from the carrier-plane to ward off enemy attack, and then returned to the mother airplane.

Rather, the air-launch system used in the research airplane program can be considered as another form of assisted take-off, similar in some respects to the two-stage rocket missile schemes.

The air-launch idea, for use with research airplanes, was first proposed in connection with flight programs of the Bell X-1. This rocket-powered research airplane was originally designed to be equipped with a turbo-pump system which would enable carrying approximately four minutes of fuel. However, the turbo-pump system was not available at the time the remainder of the airplane and propulsion system was completed, and consequently a nitrogen pressure system was substituted which cut the fuel supply to approximately two and one-half minutes.

With ground takeoff, all fuel was expended during the climb before reaching an altitude of 25,000 feet. Further, the hazards both to airplane and pilot were increased by ground takeoff.

As originally suggested by the Bell Aircraft Corporation, a mother-ship plan was adopted, using a Boeing B-29 with bomb bays modified to accommodate the X-1. Flying to an altitude of 25,000 feet or higher, the X-1 was released and thus was enabled to reach higher speeds, at the altitudes desired, than would have been possible through ground takeoff.

After the initial successes with the X-1 mother-ship program, it was decided that a Boeing B-50 should be similarly modified, thus to permit air launching from even higher altitudes than with the B-29.

To date, more than 140 of the air-launched flights, from either the Boeing B-29 or B-50, have been made in the X-1. Major Charles E. Yeager has made more such flights than any of the other pilots, who have included other Air Force officers, men from Bell Aircraft, and the NACA.

More recently, a Boeing P2B-1S, one of the few B-29-type airplanes procured by the Navy, was modified to accommodate the Douglas D-558-II Skyrocket. The Skyrocket flights, announced during 1951, which attained the highest speeds ever reached by man, and the highest altitudes, were air-launched.

In the case of the X-1, the flight program began with a series of flights to demonstrate performance guarantees. During these flights, Bell was responsible for complete operation of the B-29 mother airplane, the X-1, and the Reaction Motors rocket engine in the X-1. Upon completion of the demonstration flight tests, the B-29 was reassigned to the Air Force for maintenance and all future operation during launches, whether the X-1 was to be flown by Air Force or NACA pilots.

-2-

The X-1 may be used as a good example of the three-way partnership of industry, the military services and the NACA in the high-speed research airplane program. The airplane was designed and constructed by Bell Aircraft Corporation, with the assistance of preliminary design data from the NACA. The project was monitored by the Air Force which, as the procuring agency, had primary cognizance with regard to contractural costs and obligations.

The X-1 was instrumented by the NACA, both for the performance flights and the more comprehensive research flights which followed. Reduction of data obtained from flights has been performed by the NACA, and results made available to the industry and the military services through NACA reports.

When flown by Air Force pilots, the X-1 was used in an accelerated program to determine the maximum speed and maximum altitude for the airplane. This particular X-1 (two were originally built) has since been retired and now is at the Smithsonian Institute where it occupies a position of prominence as the first airplane ever to be flown supersonically. The second X-1 has been operated by NACA pilots in a detailed program of flight research at transonic and supersonic speeds. This airplane is still in service and continues to pay its way, producing valuable research information.

The principal differences in operation of the flight program of the D-558-II have been that Bill Bridgeman, Douglas test pilot, instead of a Navy pilot, flew the airplane to maximum speed and altitude, and also that since the airplane was delivered to NACA, the mother airplane also has been operated by NACA personnel.

In preparing for a research flight, the planning is done by a staff of research scientists at the NACA High-Speed Flight Research Station at Edwards. Each single flight is a part of a carefully programmed series, designed to provide specific aerodynamic data. While the flight planning is underway, flight-operations and instrumentation personnel are readying equipment for flight. The mother airplane is elevated on jacks, and the test airplane is rolled under and hoisted into the bomb bay.

The instruments are loaded with film, and pre-flight checks and calibrations of the instrumentation are made. The test airplane then is loaded with the special fuels and gases required: liquid oxygen, alcohol, hydrogen peroxide, nitrogen, and helium.

Then a briefing is held at which the flight plan is discussed. Attending are the research project leader, the crew of the mother airplane, the engineer-test pilot of the research airplane, and the radar crew. Decisions are made concerning the drop altitude and location, the exact headings for the test flight. The maneuvers to be executed and the information to be sought are also reviewed. Finally, the various segments of the actual flight operation are coordinated, participated in by the Air Force escort

-4-

pilot, whose mission it is to observe the test and report any malfunctions, damage or other unusual condition not visible to the pilot of the test airplane.

The takeoff is made as soon as possible after fueling to avoid excessive boiling off of the liquid oxygen. The test pilot rides in the mother airplane during takeoff and initial climb. Usually, he enters the test airplane at about 10,000 feet, to begin preparations for the drop. Continuous radio contact is used between the various units involved in a flight -- the mother airplane, the test airplane, the escort airplane, the radar tracking station, and the ground control station where personnel are stationed to assist the pilot, and to suggest last minute alterations in the flight plan because of changing weather conditions, airplane functioning, etc.

The launch is usually made about 25,000 feet, the exact altitude depending on the specific mission. The pilot fires his rocket engine as soon as practicable after the launch, and goes into his predetermined flight plan. After the fuel has been exhausted, the plane becomes a glider, but even during the glide portion of the flight, useful data are obtained. During the latter part of the glide the Air Force escort pilot stays close to the test airplane to maintain constant visual check on landing gear, flaps, etc., and also to furnish an added measure of guidance because of the limited visibility available to the test pilot. This guidance is particularly important

-5-

during the landing stage where the test pilot has difficulty in determining his exact height above the ground.

After the flight, the pre-flight procedure is essentially reversed. The airplane is checked over, the instrument film is removed and developed, and a post-flight meeting of the research scientist and the test pilot is held to discuss generally the results of the flight and to inspect the records. The records are then turned over to the data reduction group where the records are worked up in a form suitable for analysis. These data are then used by the research group for planning the next flight and accumulating the knowledge into an integrated picture. From a series of flights such as this, a complete story on some phase of the program is assembled. The results are then prepared in the form of NACA research memorandums and the data is then available for use by anyone who has need of such information.

In addition to the X-1 and two D-558-II airplanes, the NACA High-Speed Flight Research Station also operates one each of the following types: X-4, X-5, YF-92A, and D-558-I.

* * *

April 25, 1952

CONFIDENTIAL
SECURITY

REC'D SEP 2 1952

Edwards, California
August 28, 1952

From HSFRS
To   NACA

Subject: Increased thrust of the LR8-RM-6 rocket engine

Reference: NACA ltr to Langley dtd July 31, 1952; ClcW.mmt

1. As requested in the reference letter, the reasons for the desirability of operating the LR8-RM-6 rocket engine at increased thrust are submitted. There are two reasons for the need of higher thrust:

   (a). It is anticipated that the drag of the external stores that will be mounted on the D-558-II airplane will be high and will probably limit the speed obtainable at the present rated thrust to valves close to M = 1. Calculations, based on extrapolation of wind tunnel drag data, indicate that, with the stores installed, the D-558-II (No. 145) airplane will be limited to a maximum Mach number of about 1.05. An increase of 50 percent in the thrust of the rocket engine would permit the attainment of a maximum Mach number of about 1.25. The performance of the D-558-II (No. 143) airplane after its conversion to air launching and all rocket is shown on the attached figure for conditions of stores on, and off, with, and without increased thrust. The calculations at the higher Mach numbers with stores on are questionable because no drag data on stores was available at these speeds.

   (b). The utilization of a pressure suit for the pilot has markedly increased the operational altitude of the D-558-II (No. 144) to the point where increases in thrust will materially increase the Mach number attainable. Points are included on the attached figure showing the Mach number increases possible at 70,000 feet altitude from 15 percent and 50 percent increases in the thrust of the rocket engine. The increases in performance resulting from the increased thrust would considerably increase the utility of the D-558-II research vehicle.

Walter C. Williams
Chief, NACA High Speed Flight Research Station

WCW: pm

Enclosure (1)

CONFIDENTIAL
SECURITY INFORMATION

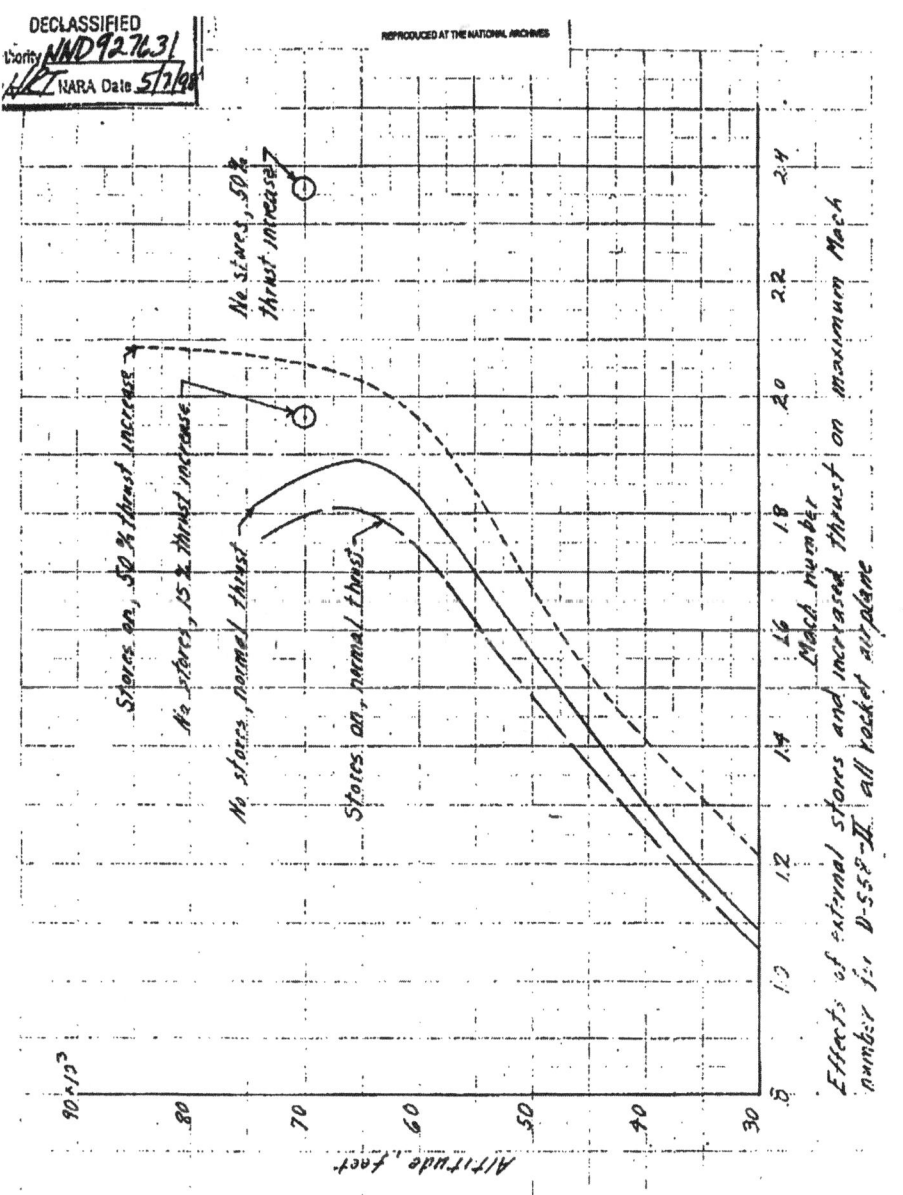

Effects of external stores and increased thrust on maximum Mach number for D-558-II all rocket airplane

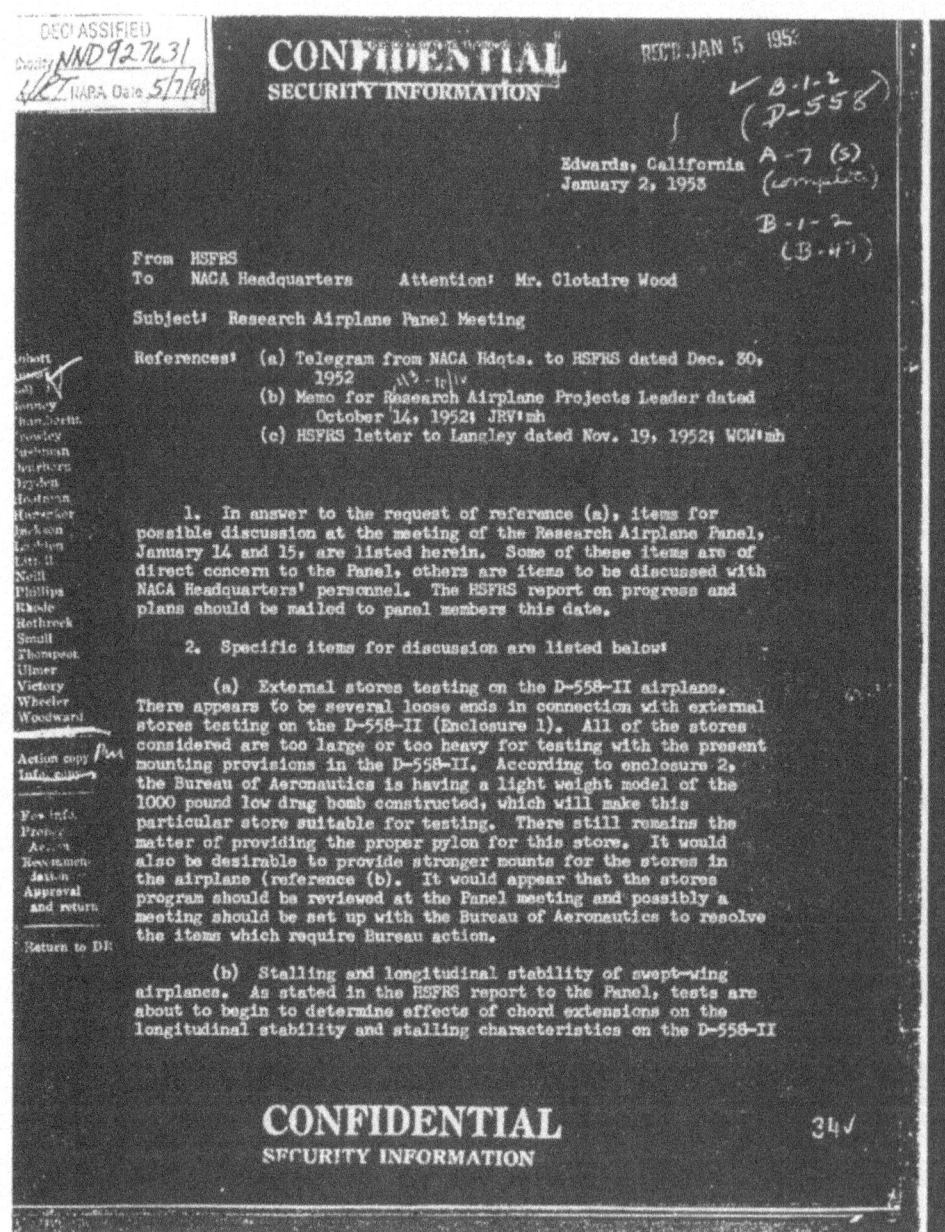

Document 38, Letter, Walter C. Williams, Chief, NACA High-Speed Flight Research Station, To: NACA Headquarters; Attention: Mr. Clotaire Wood, Subject: Research Airplane Panel Meeting, January 2, 1953

NACA 143 airplane is being used for these tests. This is also the airplane scheduled to be used for the external stores tests. It is suggested that at the Panel meeting a review be made of suggested devices for alleviation of longitudinal instability of swept-wing airplanes with a view towards resolving which are worthy of flight tests. It is not known at this time when a third D-558-II (NACA 143) will be available for flight tests. It is suggested, therefore, that if extensive flight tests of stall control devices is anticipated much of the work load be transferred to the X-5. The X-5 airplane is a very productive research vehicle in that flights can be made at a rapid rate. In the case of testing of auxiliary devices, however, considerable shop time is required to accomplish modifications. It would be desirable, therefore, to acquire the second X-5 airplane at an early date in order to minimize delays to the basic program on the X-5 airplane. At present, as reported in HSFRS's report, the second X-5 is being used by the Air Force on a Phase II evaluation and measurement of dynamic stability. It is suggested that contacts be made with the Air Force to the effect that they complete their Phase II evaluation and forego the measurement of dynamic stability, turning the airplane over to NACA. This would appear to be a reasonable request since the Air Force dynamic stability program would only duplicate what has been done and is being done by NACA on the X-5 airplane. If the Air Force program includes development of techniques, this work could better be accomplished on a production type rather than a research airplane. In addition, two other specific items of research are suggested for the X-5 airplane in the report to the Panel.

(c) Procurement of an F-86F airplane for accelerated longitudinal stability program. A request was made to NACA Headquarters through the Research Airplane Projects Leader (reference (a)) for a short term loan of an F-86F airplane for longitudinal stability evaluation. To date nothing has been heard from NACA Headquarters on this request, so perhaps some discussion is in order.

(d) Spare parts support for the B-47 airplane. To date there is no evidence of action to obtain parts on an expedited basis for support of the B-47 flight tests at Edwards (Enclosure 5). Some discussion of this subject appears in order.

3. In addition to the specific items mentioned above, it is expected that the programs and progress on the research airplanes will be reviewed. There will undoubtedly be specific

CONFIDENTIAL
SECURITY INFORMATION

discussion during this review, but it is felt that only those items mentioned above should involve any preliminary preparations.

Walter C. Williams
Chief, NACA High Speed Flight Research Station

WCW:pm

cc: Mr. H. A. Soulé

Enclosures:
(1) Memo on visit of Mr. K. J. Evans
(2) BuAero, Navy, ltr. on External Stores
(3) Memo re. spare parts for the B-47.

SENIOR SCHOOL
MARINE CORPS EDUCATIONAL CENTER
MARINE CORPS SCHOOL
QUANTICO, VIRGINIA

MEC:us
Ser: 0115

28 OCT 1953

From: Lieutenant Colonel Marion E. CARL, 06053/7333, USMC
To: Chief, Bureau of Aeronautics, Aer-AC-241

Subj: Flights in NACA D-558-2, report of

Ref: (a) BuAer ltr Serial No. 120772 dtd 9 Sep 1953

1. In compliance with reference (a) a report on subject flights is hereby submitted.

2. A total of seven (7) flights were made in the D-558-2, for a total flight time of ninety (90) minutes. All flights were air drops from NACA's B-29. All flights were chased by at least one (1) Air Force plane and usually by two (2). Mr. A. S. Crossfield, the NACA project pilot on the D-558-II, supervised all pilot check out, familiarization and indoctrination including checking the aircraft personally prior to each flight. He also flew as part of the B-29 crew on each flight and helped with hooking up the pilot within the cockpit.

3. The first two flights were in ship #145 which incorporated a J-34 jet engine in addition to the Rocket Motor (RMI). The first flight was strictly jet powered with the drop from the B-29 being made at about 30,000 feet. Duration of this flight was twenty (20) minutes. This flight consisted mainly of checking pitch up tendencies in wind up turns. Considerable loss of altitude was encountered in these maneuvers despite 100% RPM on the jet engine.

4. The second flight was planned for a launch at 30,000 feet in #145 utilizing both the jet and rocket engines. However, No. 3 engine on the B-29 developed trouble and the drop was made at about 20,000 feet. The flight plan for this flight called for a level flight run at 35,000 feet to a Vmax of 1.2 Mach. No.. As a result of this low drop a maximum IMN of about 0.92 was reached prior to exhausting all the rocket fuel. On

MEC:us
Ser: 0115

CONFIDENTIAL

Subj: Flights in NACA D-558-2, report of

the way down it was intended that a brief check on lateral, directional and longitudinal be made, but due to fumes in the cockpit this was not done. Post flight inspection revealed a hydraulic leak as being the primary reason for fumes in the cockpit. Time for this flight was nineteen (19) minutes.

5. The third scheduled flight was to be the last flight in #144, the aircraft powered by rocket only. At about 32,000 feet and a couple minutes prior to launch the pilot ran out of oxygen. A quick check revealed zero oxygen pressure on the D-558 system. As the cabin pressure was about 15,000 the face piece was opened and normal breathing resumed. The flight was aborted. All lox and fuel was jettisoned and the B-29 landed with the D-558 aboard. Post flight revealed a serious oxygen leak in the coupling attaching the suit to the aircraft's system. Since there was no check valve between this point and the pilot's emergency system carried in the back pan, this system would have been of little use had the pilot tried to utilize it. The time involved from switching from the B-29 oxygen system to depletion of the D-558 system was estimated at five (5) minutes. Thereafter the pilot carried a bail out bottle equipped with a mouth piece strapped to one leg as a third system.

6. The first successful all rocket flight was made the next day in #144. The launch was made at about 32,000 feet. The rocket chambers lighted without more than normal delay. During this climb the pilot allowed the aircraft to pick up too much speed at the start of the climb and as a result indicated speeds throughout the climb were too high. Also, considerable difficulty was encountered holding proper climbing attitude due to buffet. After the rocket fuel was exhausted the attitude was held about five seconds with the result that minimum IAS noted on the pushover was 140 mph. Full nose down elevator plus full nose down stabilizer was used on the pushover. Reaction was very sluggish and was accompanied by some lateral roll left and right that could not be completely controlled. When the nose dropped to the horizon full nose up elevator followed by full nose up stabilizer did not immediately effect the aircraft. On this and all subsequent flights the nose always went down steeper than the pilot intended and some 20,000 feet or more loss of altitude and 1.4 to 1.5 IMN was encountered before anything like full control was established. This undoubtedly could have been best avoided by shoving over immediately when the rockets quit rather than holding the nose up in an endeavor to attain the maximum altitude. Pullouts at the higher altitudes never exceeded 3 g. and even then at the lower indicated

CONFIDENTIAL

- 2 -

CONFIDENTIAL

MEC:us
Ser: 0115

Subj: Flights in NACA D-558-2, report of

speeds extreme care had to be exercised to avoid pitch up as it and buffet occurred almost simultaneously. As soon as practical a glide speed between 0.85 and 0.90 IMN was established from which point the approach and landing was routine. Touch down on the dry lake bed was normally made at an IAS of 150 mph. Maximum altitude attained on this flight was about 72,000 feet.

7. The second all rocket flight was similar in all respects to the first with almost the same results. Maximum altitude attained was about 75,000 feet.

8. The third all rocket flight started off a bit differently in that the loading of the D-558 in the B-29 was accomplished the evening prior to the flight rather than the morning of the flight. Despite unexpected delay this permitted leaving the ground at about 1000 instead of about 1300. As a result it was cooler, the lox more stable, and the B-29 was able to go higher. Launch was made at about 34,000 feet however, the D-558 dropped to 28,000 prior to heading up. This was caused by failure of the first chamber to light. Finally another chamber was touched off from which point all remaining chambers lighted normally. This time the pilot managed to stay quite close to the optimum climb schedule which was just below the buffet range. As a result no buffet was encountered and control in the climb was more positive. It was noted that the rockets quit at about 75,000 feet indicated altitude and that 80,000 feet indicated was reached on the pushover. Minimum indicated speed on the pushover was 150 mph IAS. An IMN of 1.5 was attained prior to recovery from the dive.

9. The fourth flight was made for the purpose of attaining maximum speed. As on the third flight the B-29 was loaded the night before and take off was made at about 0840. Launch was made at about 32,000 and the same climb schedule was maintained to about 58,000 feet at which point a pushover was made. During this pushover which was about 0.2 to 0.4 g. a violent lateral oscillation started which the pilot was unable to control. The rudder was locked in the neutral position prior to launch and an endeavor was made to stop the oscillation with the ailerons but they seemed to be quite ineffective. The aircraft rolled left and right to about 75 deg and finally at about 1.5 IMN the rockets quit presumably due to unporting the fuel as about 200 rocket seconds were left.

10. The fifth and last flight on #144 was made two days later. The B-29 took off at about 1100. The launch was made at about 31,000 feet and this time pushover was made at 55,000 feet at 0.4 g. During the pushover a maximum indicated altitude of

CONFIDENTIAL

- 3 -

MEC:us
Ser: 0115

CONFIDENTIAL

Subj: Flights in NACA D-558-2, report of

68,000 was noted. Up to this point it was a good flight even the lateral oscillations were being well controlled by energetic use of the ailerons. However, the pilot at this point permitted the dive to become too steep with the result that the aircraft descended too low too soon precluding optimum speed. A true air speed of about 1140 mph was reached at about 48,000 feet. There was evidence also that not quite all of the fuel was utilized. For this flight, time from launch to landing on the dry lake bed was nine (9) minutes.

11. Mechanics of each flight as far as the pilot was concerned from B-29 take off to launch off the D-558 was always the same. Immediately after take off of the B-29 the pilot shed his clothes and donned special underwear, then with the help of the representatives from AMEL and the David Clarke Company climbed into the pressure suit. At about 7,000 feet the pilot was ready and started entry into the D-558-II. At about 10,000 feet everything was connected, the canopy closed and the ready to launch light turned on. From this point until five minutes prior to launch the pilot was merely along for the ride and did almost nothing. At five minutes prior to launch the pilot started his check off with Mr. Crossfield on the intercom and performed all steps as required up to the drop. At drop the only remaining item was lighting off the rocket chambers. No. 3 chamber was always touched off first, then Nos. 4, 1 and 2. The first two were touched off as fast as chamber pressures permitted or within about ten seconds after launch. An IAS of about 320 was then attained the nose started up and the remaining two chambers fired. On the third flight in a #144 this was modified when No. 3 would not fire, to going on to No. 4 then coming back to No. 3 at the end.

12. The hardest part of each flight was that of maintaining the proper attitude in the climb. There was no attitude gyro or artificial horizon in the cockpit and the pilot could see neither ground nor horizon. Movement within the cockpit was extremely limited. It was possible for the pilot to turn his head about 30 deg left and right. The top of his head touched the canopy, and shins were up against the instrument panel so that only the tip of the toes touched the pedals. Because of the back pan the pilot was forced so far forward that aft movement of the stick was limited to something less than full up elevator. It was obvious that if the suit inflated, control would have been marginal and could probably have been maintained only by use of the electric powered stabilizer down to below 35,000 feet

- 4 -

CONFIDENTIAL

CONFIDENTIAL

MEC:us
Ser: 0115

Subj: Flights in NACA D-558-2, report of

when the suit would have deflated. Many of the engine instruments were hidden by the pilot's legs and it was very difficult to move the legs enough to see them due to the cramped position and limited space.

13. Due to the limited time spent in the aircraft and various problems encountered during this time such as establishing positive control in the pullout, setting up the proper glide speed, and finally the proper pattern for a dead stick landing, no detailed evaluation of the handling characteristics was made. In general it was felt that directional stability was satisfactory and rudder control adequate for such an airplane. Lateral stability was also satisfactory--infact, dihedral effect was quite strong and particularly noticeable in the approach to a landing. Lateral roll left and right encountered at high altitude appeared to be more a function of low acceleration and was particularly noticeable anytime zero g was approached. Full and decisive use of the ailerons were required to properly control the aircraft once this oscillation started. Application of positive g damped the oscillation considerably and no tendency for oscillation was encountered above 1.0. g. Longitudinal force stability was slightly positive and actually appeared to be negative near and in the buffet range. As the accelerated stall was approached there was a strong tendency to pitch up which sometimes required full forward stick to counteract. At high altitude and high speed this can become a most uncomfortable and vicious maneuver if corrective action is not promptly taken at first evidence of pitch up. At all normal speeds and conditions control harmony and effectiveness was quite good.

14. The pressure suit utilized on the last five flights was a full pressure type manufactured by the David Clark Company of Worcester, Massachusetts. This suit was about two years in the making. The pilot was able to open and close the mouth piece and remove and replace the face lense. It is not possible for the pilot to don or remove the suit unaided. In fact, the pilot carried a knife for the express purpose of cutting his way out should the circumstance arise in which it was necessary to get out aided.

15. After the suit was properly fitted, two and a half days were spent at AMEL in indoctrination to the wearing and functioning of the suit. As the suit fitted almost skin tight the first problem was fighting off claustrophobia which turned out

CONFIDENTIAL

CONFIDENTIAL

MEC:us
Ser: 0115

Subj: Flights in NACA D-558-2, report of

to be quite an item when the pilot became too warm. After the first couple times in the suit this was never again of any consequence. In addition to the present schedule for indoctrination which includes both link trainer time with suit inflated and deflated, a run to 80,000 feet in the chamber, and couple explosive decompressions, it is felt that a flight in a TO-2 with the pilot flying with suit inflated for a short time would have been of definite value.

16. At present one of the biggest problems is the control of body heat. The pilot is either too hot or too cold. Normally this turned out to be too hot - to the point that a certain amount of perspiration was poured out at the same time as the pilot. To a certain extent this could be alleviated by better ventilation. Ventilation around the head and face as well as the arms and legs should be provided. Increased mobility is necessary. At this time the pilot is severely limited particularly when the suit is inflated. With the suit deflated it is not possible to see most of the controls and gages that are located on the side consoles aft of the midway point. Neither can the pilot see his own safety belt. When not restrained by the canopy as was the case in the D-558-II the pilot can still turn his head only about 45 deg. It is easier to tilt the head up than it is to tilt it down. Movement of the arms and legs is considerably restricted and it is particularly difficult to climb into and out of the cockpit. The pilot was continually annoyed by inability to grasp and feel things properly. The gloves were too slick and too cumbersome. Increase vision through the face plate would be desirable - probably by increasing area of the glass and reducing width of metal rims. Tinting the face glass would be desirable to reduce glare. The weight and thickness of the back pan and pack is excessive. The pan alone without the chute was about 25 lbs. and total thickness of pan and chute was about 5 inches.

17. Conclusions and recommendations.

    a. No detailed conclusions or recommendations on the aircraft are advanced. The pilot saw but little of the instrumented data taken on the flights - however, it should now be available from NACA. In general the pilot formed a high opinion not only in respect to the reliability of the airplane but also of the program as conducted and all NACA personnel the pilot came in contact with.

- 6 -

CONFIDENTIAL

CONFIDENTIAL

Subj: Flights in NACA D-558-2, report of

    b. In respect to the pressure suit the following recommendations are made.

        (1) Increase suit ventilation and establish a method of positive heat control.

        (2) Increase mobility of the suit in general.

        (3) Re-design the suit to permit a pilot to don and remove the suit unaided.

        (4) Decrease weight and size of the back pack and chute.

        (5) Increase the coefficient of friction on the face of the gloves.

        (6) Increase mobility of the fingers.

        (7) Increase vision.

Conclusions:

It is concluded that the suit is satisfactory for continued flight use as an interim measure pending improvements, and that it is the best suit for high altitude research flights now available.

                                            MARION E. CARL

CONFIDENTIAL

Edwards, California
January 25, 1954

MEMORANDUM for NACA Headquarters

Subject: Rocket nozzle extensions used on the LR8 engine for the D-558-II, No. 144 airplane.

References: (a) Letter from HSFRS to RAPL dated 8-19-52, FMD:mem.
(b) Letter from RAPL to NACA dated 8-20-52, HAS:jcl.
(c) Letter from NACA to BuAero dated 8-25-52.
(d) Letter from NACA to BuAero dated 9-30-52, Clow:pap.
(e) Letter from NACA to HSFRS dated 12-15-53, Clow:dlf.
(f) Letter from HSFRS to Chief of Research dated 2-21-52, DRB:mem.
(g) Letter from NACA to HSFRS dated 10-10-52, BEG:eh.

1. Reference letters (a), (b), (c), (d), and (e), deal with a request for nozzle extensions for the LR8 rocket engine used in the D-558-II airplane. These extensions were requested primarily to reduce the abrupt pressure change at the nozzle exit with the possibility of eliminating the adverse rudder hinge-moment characteristics encountered during high altitude flights at supersonic Mach numbers.

2. The nozzle extensions were first used in September, 1953, and a great improvement in the rudder hinge moments was apparent. Without the nozzle extensions, the rudder hinge-moment slope $Ch_{\alpha r}$ had a large positive value whenever the rocket engine was operating during high altitude flights with the maximum value occurring at a Mach number in the vicinity of 1.5. With the nozzle extensions, which reduced the exit pressure ratio at an altitude of 60,000 feet from 13 to 5, $Ch_{\alpha r}$ remained negative.

3. An additional benefit was derived from the use of the extensions in that at altitudes above 16,000 feet there was an increase in thrust which amounted to about 500 pounds at an altitude of 70,000 feet. Figure 1 shows the effects of the extensions on the thrust. The loss in thrust at the lower altitudes is not significant for the D-558-II because it is not the normal powered flight range.

4. Reference letters (f) and (g) were concerned with the ability to determine nozzle coefficients during ground thrust stand runs when the gasses are overexpanded as would be the case when nozzle extensions are used. Therefore, the value of the nozzle coefficient was determined by three different means: (1) Theoretically, using the equation

$$c_n = \sqrt{\frac{2\gamma^2}{\gamma-1}\left(\frac{2}{\gamma+1}\right)^{\frac{\gamma+1}{\gamma-1}}\left[1-\left(\frac{p_e}{p_c}\right)^{\frac{\gamma-1}{\gamma}}\right]}$$

where γ is the ratio of specific heats, p_e is the nozzle exit pressure, and p_c is the combustion chamber pressure. (2) Experimentally by making a ground run on the Edwards Air Force Base thrust stand, and (3) Experimentally in flight at altitudes of about 60,000 feet by noting changes between power-on and power-off conditions. The following values were obtained:

	Theoretical	Ground Run	Flight
Without nozzle extension	1.386	1.354	1.384
With nozzle extension	1.546	1.513	1.535

It is apparent that with these particular extensions which expand the gas to 5 pounds per square inch, little or no separation occurs and that the nozzle coefficients can be determined during ground runs.

5. There has been a certain amount of deterioration of the nozzle in flight as shown in figure 2. In the cases of burnout in flight, there was no effect of the flame on the base plate or other parts of the airplane. The burnouts have occurred in some definite relationship to the nozzle injector pattern. Experiments are now in progress to determine if inconel or some other metal would be more suitable for the nozzle extensions.

Donald R. Bellman
Aeronautical Research Scientist

DRB:mh

cc: Langley
    Ames
    Lewis

CONFIDENTIAL

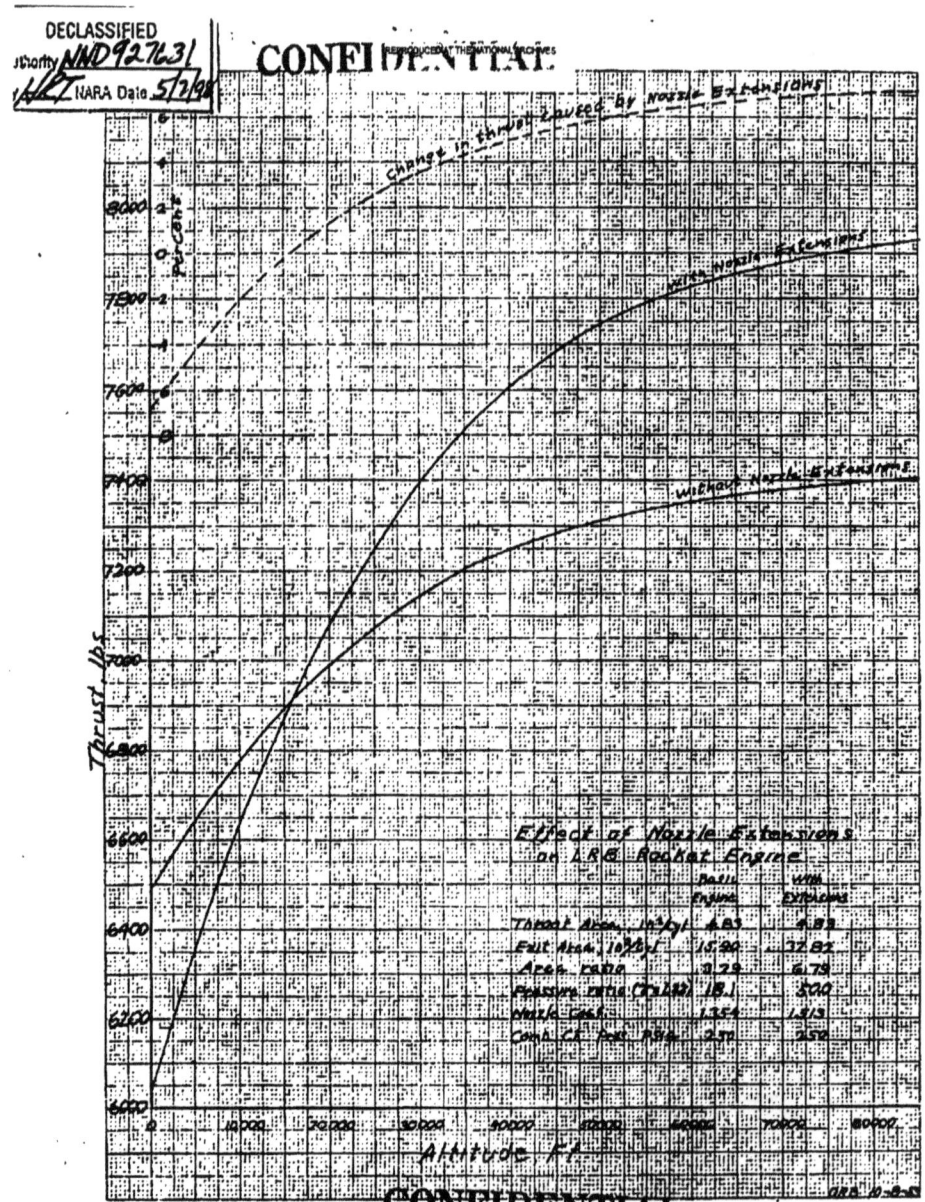

February 3, 1954

From   : National Advisory Committee for Aeronautics
To   : Chief, Bureau of Aeronautics
            Department of the Navy
            Washington 25, D. C.

            Attention: Mr. Oscar Bessio
Subject   : Results of use of rocket nozzle extensions for the LR8 engine on D-558-II airplane

References: (a) BuAer ltr to NACA dtd 23 July 1952, Aer-SI-531
            (b) NACA ltr to BuAer dtd 27 Dec. 1951, LES:bbn

1. The use of uncooled nozzle extensions in lieu of redesigned cylinder nozzles to alleviate rudder oscillations due to shock waves from rocket jet expansion was suggested by the Bureau of Aeronautics in the referenced letter. The nozzle extensions were subsequently built and provided to the NACA.

2. Use of the uncooled nozzle extensions has resulted in considerable improvement in the rudder hinge moment characteristics occurring during high altitude flights of the D-558-II at supersonic Mach numbers. It is believed that the results will be of interest to the Bureau of Aeronautics in connection with the suggestions of the NACA in Reference (b) that rocket cylinders be designed for correct expansion at altitudes of 30,000 to 40,000 feet. For the use of the Bureau of Aeronautics, the results obtained on the D-558-II have been summarized in a brief memorandum, a copy of which is enclosed. It would be appreciated if the enclosure is also brought to the attention of the Aircraft Division of the Bureau of Aeronautics.

CONFIDENTIAL

Chief, Bureau of Aeronautics - 2 - February 3, 1954
Department of the Navy

    3. The assistance provided to the NACA by the Bureau of Aeronautics in furnishing the subject nozzle extensions is appreciated.

                            Ira H. Abbott
                            Assistant Director
                            for Research

Enclosure
Cy Memo fr. HSFRS dtd 25 Jan 54
w/encls.

CloW:dlf

CC: HSFRS (Ref. HSFRS ltr dtd 25 Jan. 54, mh)

CONFIDENTIAL

Edwards, California
April 29, 1954

MEMORANDUM to Mr. Bonney

1. With reference to your letter of April 19, 1954, I will attempt to amplify the research piloting experience of Joe Walker and Stan Butchart.

2. Walker has had piloting assignments on the D-558-I, X-1, X-4, X-5, and B-47; while Butchart has been with the D-558-I, X-4, X-5, D-558-II (jet powered) and B-47. The flight programs for these airplanes have been concerned with Handling Qualities, Aerodynamic Loads and Performance. In addition, the program on the B-47 has been concerned with determining the effects of flexibility on aerodynamic loads and stability. The programs on the B-47, X-4, X-5, D-558-I and D-558-II (jet powered) have been in the subsonic and transonic speed range, while, of course, the X-1 is in the transonic and supersonic range.

3. I would also like to point out that Joe Walker's piloting experience at Lewis was concerned with icing, ram jet flight studies, and high performance fuel tests. Incidentally, Eastern Airlines in their letter of March 13, 1951 to Dr. Dryden was complimentary of NACA TN 2300, of which Joe Walker was co-author.

4. I'm sending some shots of Walker for you to look over. If they are not what you want, please let us know and we'll get additional ones made. Stan Butchart went on leave before Hodgepeth could get some shots. I will follow up on this and forward them on to you.

5. I'm attaching a copy of Marvin Mills' column of April 25, which may be of interest to you.

Marion I. Kent
Administrative Officer

MIK:lmp

*Hope to see you when we open up the new building June 26th.
M.*

Edwards, California
July 29, 1954

From HSFS
To NACA Headquarters

Subject: Low temperature difficulties with hydrogen peroxide in model D-558-II aircraft

Reference: Memo NACA Hdqtrs to HSFS, CloW:dlf, July 23, 1954, with inclosure from BuAer, July 6, 1954, B.F.Coffman

1. Regarding the problem of peroxide freezing in the lines of the D-558-II airplane, between the peroxide tank and the preheat by-pass orifice, this Station does not consider the situation of sufficient importance to justify the expense involved with installing electrical heating blankets for the following reasons:

   a. During approximately 7 years of operation of the D-558-II airplanes only once have the conditions been such that this problem presented itself. These conditions were: cold peroxide from the storage barrel, ambient ground temperature about 5°F and the aircraft was allowed to stand for a long period of time on the ground, after the liquid oxygen tank had been filled.

   b. This Station took remedial action on subsequent flights by heating the peroxide to approximately 50°F prior to filling the aircraft and shortened the delay between liquid oxygen filling and take-off.

De E. Beeler
Acting Chief, High Speed Flight Station

JRV:rmf

REC'D NACA
AUG 6 - 1954

Document 43, Letter, De E. Beeler, Acting Chief, [NACA] High-Speed Flight Station, To: NACA Headquarters, Subject: Low temperature difficulties with hydrogen peroxide in the model D-558-II airplane, July 29, 1954

DEPARTM
BUREAU
WASHINGTON 25, D. C.

Aer-SI-53

C O N F I D E N T I A L

2 JUN 1954
010866

From: Chief, Bureau of Aeronautics
To: National Advisory Committee for Aeronautics, 1724 F St., N. W. Washington 25, D. C.

Subj: Improved version of LR8 liquid rocket engine for use in D558-2 airplanes; availability of

Ref: (a) NACA ltr dtd 11 April 1952

Encl: (1) Tentative General Specifications, RMI Model Designation TR 114, 6000 lb. and 8000 lb. ratings.

1. The development of an improved version of the LR8-RM-6 liquid rocket engine currently used in the D558-2 airplane is approaching completion. This work is being performed at Reaction Motors, Inc. under Bureau of Aeronautics contract NOas 52-802-c. It is expected that new engine parts and assemblies will be ready during the latter part of this year for incorporation in the flight engines to convert them to the new configuration.

2. The LR8 redevelopment program is based on field operating experience and attempts to improve the performance, reliability, reproducibility and maintenance of the rocket power plant. Changes in the design of a number of components have been made, notably the thrust chambers, igniters, propellant valves and the turbopump governors.

3. The design and operating characteristics of the improved LR8 engine are described in enclosure (1). It will be noted that this engine can be rated at either 6000 lbs. or 8000 lbs. of thrust. This is possible since the new engine components have been stress designed, proportioned and are being tested at the 8000 lb. level. It is believed that this feature of the engine will be of considerable value to the flight program.

4. In reference (a) the NACA indicated that thrust in excess of 6000 lbs. would be of little benefit to the flight program unless rocket thrust duration was increased. It will be noted from enclosure (1) that operation at 8000 lbs. is accompanied by an appreciable improvement in specific impulse. Therefore, the higher thrust rating of the improved power plant should be of interest.

5. The comments of the NACA on the thrust level desired for future D558-2 airplane operations should be furnished to this bureau so that final adjustments can be made to the engine. The currently rated 8000 lb. engine can be made to operate at 6000 lbs. by either decreasing pump speed or trimming the pump impellers, an operation that can be accomplished in the field.

Copy to:
RMI, Rockaway, N.J.

C O N F I D E N T I A L

B. F. COFFMAN
By direction

Turborocket Engine

RMI Model Designation TR 114

(6000 # Thrust)

Dry Weight Thrust Chamber Assembly (4 Thrust Chambers)	243# (max.)
Dry Weight Turbopump Assembly	97# (max.)
Total Engine Weight	340# (max.)
Number of Thrust Chambers	4
Sea Level Thrust Rating, lbs.	
1. Individual Chamber	1500 (min.)
2. Complete Engine	6000 ± 5% -0%
Thrust at 50,000 Feet, lbs.	
1. Complete Engine	7000
Thrust at 50,000 Feet with 6.7 area ratio through use of Nozzle Extensions, lbs.	
1. Complete Engine	7550
Design Oxidizer-Fuel Ratio (O/F)	1.16 ± 3%
Specific Impulse (design O/F)	192 sec. (min.)
Propellants:	
1. Fuel	Alcohol/MIL-A-6091 - water mixture rocket fuel mixed to a standard specific gravity of .850 ± .010 at 60°F.
2. Oxidizer	Liquid oxygen/AN-O-1C, Grade B, Type II
3. Pressurizing gas	Nitrogen/MIL-N-6011
4. Pump drive	90% hydrogen peroxide
Propellant Consumption	
1. Fuel	14.5 P.P.S.
2. Oxidizer	17 P.P.S.
3. Pump Drive	0.3 P.P.S. (1 chamber) to 0.55 P.P.S. (4 chambers)
Turbopump Suction Pressures	
1. Oxidizer	15 psi NPSH
2. Fuel	13 psi NPSH
Manifold Pressures	
1. Oxidizer**	320 psia to 350 psia
2. Fuel**	350 psia to 370 psia

**Permissible operating pressure differential; 10 to 40 psi

CONFIDENTIAL

Tentative General Specifications
Turborocket Engine
RMI Model Designation TR 114
(6000 # Thrust)

Pump Drive Inlet Pressure	415 to 440 psia
Inert Gas Bleed	
1. Bleed During Run	6.4 ft^3/min. S.T.P.
Inert Gas Inlet Pressure	415 to 440 psia
Electrical Requirements	20-28 V.D.C. 14A (max.)
Type of Ignition	Ethyl alcohol-water mix, oxygen fed igniter with high tension spark plug.
Thrust Chamber Cooling	Regenerative (fuel)
Feed System	H_2O_2 gas generator driven turbopump
Thrust to Engine Weight Ratio	17
Controllability	Engine is capable of operation with chambers started or stopped individually in any sequence by manual switch control.
Starting and Operating Attitude	Vertical (thrust vector up) to 30° down thrust vector
Design Acceleration Loads	10 g downward along vertical center line, 5 g fore and aft along horizontal plane, 5 g acting upward, 5 g acting sideward in either direction

145

July 23, 1954

From: National Advisory Committee for Aeronautics

To: Chief, Bureau of Aeronautics
Department of the Navy
Washington 25, D. C.

Subject: Improved version of LR8 liquid rocket engine for use in D-558-II airplanes; availability of

Reference: BuAer ltr to NACA dtd 2 June 54, Aer-SI-53
Ser 010866

    1. In reply to reference letter, it will be desirable to use the 8,000 lb. thrust version of the LR8 engine in the D-558-II airplanes, since the improvement in specific impulse will increase flight time available at high Mach numbers. The increased thrust will improve the flight performance slightly and will increase the flexibility of the engine by permitting greater thrust variations.

    2. The exact amount of improvement in performance cannot be determined since the ratings given in the reference letter were only for the 6000 pounds thrust engine. If similar data are available for the 8000 pounds thrust engine, it will be appreciated if Bureau of Aeronautics can provide them directly to the NACA High-Speed Flight Station at Edwards Air Force Base, Attention: Mr. W. C. Williams.

    3. It appears that the change to the improved engine can be accomplished gradually during normal overhaul periods. Certain of the improved parts are interchangeable with existing parts and two of these, the turbine governor and the gas generator, offer an immediate benefit by greatly reducing maintenance time. Therefore, it is suggested that these two items be sent to Edwards as soon as they have been checked on complete engines and are available for use on present engines.

Ira H. Abbott

CloW:dlf

CC: HSFS

Ira H. Abbott
Assistant Director
for Research

CONFIDENTIAL

Document 45, Letter, Ira H. Abbott, [NACA] Assistant Director for Research, To: Chief, Bureau of Aeronautics; Department of the Navy, Subject: Improved version of the LR8 liquid rocket engine for use in the D-558-II airplane, July 23, 1954

DEPARTMENT OF THE NAVY
BUREAU OF AERONAUTICS
WASHINGTON 25, D. C.

IN REPLY REFER TO B-3-2
                     B-1-2

Aer-SI-532

CONFIDENTIAL   D-558
                LR8 Engine

11 AUG 1954

From: Chief, Bureau of Aeronautics
To: National Advisory Committee for Aeronautics
      1724 F Street, N. W.
      Washington 25, D. C.

Subj: Improved Version of LR8 Liquid Rocket Engine for Use in
      D558-2 Airplanes; availability of

Ref: (a) NACA ltr dtd 23 July 1954
      (b) Bu Aer ltr Aer-SI-53 ser 010866 dtd 2 June 1954

Encl: (1) Tentative General Specification, RMI Model Designation
      TR114, 8000 lb. rating

1. Reference (a) states that the specification for the 8000 lbs. engine version was not received with reference (b) by the National Advisory Committee for Aeronautics. Since it appears that this specification may have become inadvertently detached from reference (b), enclosure (1) is forwarded for your information.

B. F. COFFMAN
By direction

CONFIDENTIAL

Document 46, Letter, B.F. Coffman, Bureau of Aeronautics, To: National Advisory Committee for Aeronautics, Subject: Improved version of the LR8 liquid rocket engine for use in the D-558-II airplane, August 11, 1954

Page 1 of 2

TENTATIVE GENERAL SPECIFICATIONS

Turborocket Engine

RMI Model Designation TR 114

(8000 # Thrust)

Dry Weight Thrust Chamber Assembly (4 Thrust Chambers)	243# (max.)
Dry Weight Turbopump Assembly	97# (max.)
Total Engine Weight	340# (max.)
Number of Thrust Chambers	4
Sea Level Thrust Ratings, lbs.	
1. Individual Chamber	2000 ± 2-1/2%
2. Complete Engine	8000 ± 2-1/2%
Thrust at 50,000 Feet, lbs.	
1. Complete Engine	8770 ± 2-1/2%
Thrust at 50,000 Feet with 6.7 area ratio through use of Nozzle Extensions, lbs.	
1. Complete Engine	9300 ± 2-1/2%
Design Oxidizer-Fuel Ratio (O/F)	1.16 ± 3%
Specific Impulse (design O/F)	202 sec. (min..)
Propellants:	
1. Fuel	Alcohol/MIL-A-6091 - water mixture rocket fuel mixed to a standard specific gravity of .850 ± .010 at 60°F.
2. Oxidizer	Liquid oxygen/AN-O-1C, Grade B, Type II
3. Pressurizing gas	Nitrogen/MIL-N-6011
4. Pump drive	90% hydrogen peroxide
Propellant Consumption	
1. Fuel	18.3 P.P.S.
2. Oxidizer	21.3 P.P.S.
3. Pump Drive	0.35 P.P.S.(1 chamber) to 0.65 P.P.S.(4 chambers)
Turbopump Suction Pressures	
1. Oxidizer	15 PSI NPSH
2. Fuel	13 psi NPSH
Manifold Pressures	
1. Oxidizer **	445 psia to 475 psia
2. Fuel **	475 psia to 495 psia

**Permissible operating pressure differential: 10 to 40 psi

CONFIDENTIAL

Tentative General Specifications

Turborocket Engine

RMI Model Designation TR 114

(8000 # Thrust)

Pump Drive Inlet Pressure	415 to 440 psia
Inert Gas Bleed	
1. Bleed During Run	7.7 ft^3/ min. S.T.P.
Inert Gas Inlet Pressure	500 to 525 psia
Electrical Requirements	20-28 V.D.C. 14 A (max.)
Type of Ignition	Ethyl alcohol-water mix, oxygen fed igniter with high tension spark plug
Thrust Chamber Cooling	Regenerative (fuel)
Feed System	H_2O_2 gas generator driven turbopump
Thrust to Engine Weight Ratio	23.5
Controllability	Engine is capable of operation with chambers started or stopped individually in any sequence by manual switch control
Starting and Operating Attitude	Vertical (thrust vector up) to 30° down thrust vector
Design Acceleration Loads	10 g downward along vertical center line, 5 g fore and aft along horizontal plane, 5 g acting upward, 5 g acting sideward in either direction.

August 25, 1955

From: National Advisory Committee for Aeronautics

To: Deputy Chief of Staff/Development
United States Air Force
Washington 25, D. C.

Subject: Supply Support for B-29, NACA Serial No. 137, based at NACA High-Speed Flight Station, Edwards, California.

The subject airplane is the mother ship for the D-558-II research airplanes, and was donated to the NACA by the Navy, Bureau of Aeronautics, under the provisions of Public Law 672, 81st Congress. At the time of its transfer this was the only B-29 still in active use by the Navy. Supply support for the airplane was discontinued by the Navy recently when their supply of spare parts became exhausted.

Maintenance spares on a nonreimbursable basis are needed from time to time for the subject aircraft. The D-558-II airplanes are being used in research on carriage of external stores at supersonic speeds, problems of lateral stability and control at transonic and supersonic speeds, and magnitude and distribution of air loads at supersonic speeds. Although these research studies being conducted with this airplane have not been directly requested by the Air Force, it is believed that the program is of sufficient direct interest to the Air Force to warrant the issuance of spare parts by the Air Force on a nonreimbursable basis under Paragraph 16a(1), Volume II, AFM 67-1. It has been estimated that the value of these parts, including engine changes, would be approximately $150,000.00 per year.

The cooperation and assistance of the USAF in this matter will be appreciated.

John W. Crowley
Associate Director for Research

REC
CRY:dcw

Document 47, Letter, John W. Crowley, [NACA] Associate Director for Research, To: Deputy Chief of Staff/Development, United States Air Force, Subject: Supply support for the B-29, NACA Serial No. 137, based at the NACA High-Speed Flight Station, Edwards, Calif., August 25, 1955

CONFIDENTIAL

June 17, 1957

Chief,
Bureau of Aeronautics
Washington 25, D. C.

Via: U. S. Navy Liaison Office
Air Force Flight Test Center
Edwards Air Force Base, California

Subject: Completion of D-558-II Research Program

References: (a) Aer-PP-73, 06728, dated 22 May 1957
(b) NACA/F11, Ser 010, 24 May 1957

Dear Sir:

    The two reference letters inquire as to the disposition of the LR8-RM-8 engine, Serial Number 53, and the spare parts and related equipment. This engine represents an improvement over the earlier LR8 rocket engines in that the nominal thrust has been increased from 6000 to 8000 pounds and the specific impulse has been increased from about 190 to 207 seconds. In addition to the increased performance there have been other modifications to improve the operation and reliability of the engine.

    The -8 engine was originally intended for the D-558-II airplanes, but flight tests on these airplanes have now been completed, and it is felt that the small increase in performance that would result from the use of the -8 engine would not warrant extending the flight tests because of the large amount of manpower involved in these tests. The possibility of using an -8 engine in the X-1E airplane has been considered and calculations show that the maximum attainable Mach number could be increased 0.3 by such a change. However, the cylinder configuration of the -8 engine is set for the D-558-II airplane and would have to be changed if the engine were to be put into the X-1E airplane. Such a change would be rather extensive and time consuming. Furthermore, the No. 53 engine has not yet been qualified for flight which

CONFIDENTIAL

CONFIDENTIAL

might further delay its use. Considering all the above items, it is felt that this Station will not be able to use the LR8-RM-8 rocket engine.

Some of the parts of the -8 engine such as the pump control valve and the catalyst bed are interchangeable with those on other LR8 engines now in use on the X-1E and X-1B airplanes and represent an improvement over the earlier parts. It would be appreciated if these parts could be made available to these projects.

Since the NACA D-558-II program has been concluded, this Station has no further need for LR8-RM-6 rocket engines, or related equipment. We would, however, like to retain all LR8-RM-6 spare parts that are usable on the LR11 engine to support the X-1 airplanes.

Very truly yours

Joseph R. Vensel
Acting Chief, NACA High-Speed Flight Station

DRB:pm

DEB

cc: NACA Headquarters (3)
    w/cys Ref. (b) (3)

CONFIDENTIAL

Index

A-4, 10
Abbott, Ira H., 109, 139-140, 146
Aerodynamic center, 52
Aerodynamische Versuchsanstalt, 29
Air Force Flight Test Center, 1
Air-launch technique, 118-123
Angle of attack, 24 n
Ankenbruck, Herman O., 34, 54 n, 77-85
Apt, Milburn, 27, 54
Armstrong, Neil A., 34-35, 40, 43-46
Army Air Corps, 21, 22
Army Air Forces, xii, 3, 7, 22, 58
Artz, James H., 51, 53
Aspect ratio, defined, 4 n

B-29, 18-19 ill., 36, 40-41, 50, 119, 150, see also P2B-1S
B-47, 10, 127
B-50, 119
B-52, 10, 42
Beeler, De E., 142
Bell Aircraft Corporation, 4, 22, 57, 119-120
Bellman, Donald R., 67-76, 136-137
Bf 109, Messerschmitt fighter, 2
Boeing Airplane Co., 10, 38
Boeing 707, 10
Boeing 757, 10
Boeing 767, 10
Boeing 777, 10
Boyd, Albert, 7 n, 40, 60
Bridgeman, Bill, 32-33, 40, 53-54, 60, 121
Bush, George, 36
Butchart, Stanley P., x ill., xii, 34-46, 42 ill., 50, 60, 141

C-45, 39-40
C-47, 19
Caldwell, Turner, 7, 60
Carl, Marion, 7, 34, 40, 51, 60, 113-114, 129-135
Cassidy, W. B., 113
Century series of fighters, xi,
Champine, Gloria, xiii, 11
Champine, Robert A., x ill., xii, 8, 11-13, 14-23, 15 ill., 17 ill., 19 ill., 21 ill., 31, 38-39, 50, 60
Chance Vought, 26-27
Chord, defined, 4 n
Clark, David M., 51

Clark, J. R., 26
Coffman, B. F., 143, 147
Color for research airplanes, 116
Compressibility, xii, 3,
Conlon, Emerson, 4
Conrad, John, 52
Coupling dynamics, see High-speed instability
Crossfield, A. Scott, x ill., xii, 1, 10, 12 ill., 22, 32, 34, 35-41, 43, 46-56, 60, 109-113, 129
Crowley, J. W., 110-111, 150

D-558-1 Skystreak, xi-xii, 2-11, 20, 21-22 ill., 24-25, 28, 38-39, 61-62, 65, 73-76
   bureau and tail numbers, 5n
   TG-180 (J35) engine, 6, 8n
   first flight, xii, 7
   first NACA flight, 20
   stalls, 25
   ultimate load, 30
   cockpit, 31, 38
   all-movable horizontal stabilizer, 49 n, 56-57
   numbering system, 58
   drawings and data, 61-62
   elevator vibration, 67-72
   tail buffeting, 73, 75
   aileron roll data, 74-75
   dynamic stability, 75-76
   loads, 87
D-558-2 Skyrocket, xii, 4-5, 7, 12, 13, 20, 24-25, 28-36, 29 ill., 33 ill., 36 ill., 39, 40, 41 ill.-42 ill., 48-56, 61-62, 77, 84-109, 129-135
   bureau and tail numbers, 5 n
   ultimate load, 30
   cockpit, 31
   first flight, 31
   modification to air launch, 31, 65
   speed and altitude records, 32-34, 51, 54-56
   airplane that wrote the book, 48
   swept wings and handling qualities, 48 ff.
   air launching, 49, 121-123
   all-movable horizontal stabilizer, 49 n, 56-57
   turn-and-bank indicator, 50 n
   comparison of swept wing with X-1's straight wing, 52
   electrical power, 52
   drogue chute, 53, 101

nozzle extensions, 55, 136-140
other modifications for Mach 2 flight, 55
drawings and data, 61-62
stability, 78-79, 82, 83
buffet, 79
drag, 79
aerodynamic heating, 79
hydrogen peroxide and, 142
D-558-3 (never built), 58
David Clark Co., 51, 113
DeGraff, William, 42 ill.
Delavan, Charles, 11, 60
Dihedral, 54
Divergence, 54
Donaldson, E. M., 7
Douglas Aircraft Co., 1, 4, 7, 8, 9, 29, 32-33, 58, 65, 67, 69, 104
Dryden Flight Research Center, xii, see also High-Speed Flight Research Station, High-Speed Flight Station, Muroc Flight Test Unit (all earlier names)
Dryden, Hugh L., 32, 34, 54
Dunne, John, 57
Dutch Roll, 31
Dynamic stability and instability, 54 n

Edwards Air Force Base, see Muroc Army Airfield
 South Base, 49
 runway at South base, 53
Ejection system and procedures, 39, 45
Engel, Richard, Maj. Gen., 1
Everest, Frank 'Pete,' 32, 40, 60
Explorer II (balloon), 33
External stores, 35, 36 ill., 105, 124, 126

F3F, 31
F7U Cutlass, 26-27
F-84, 37
F-86, 12, 13, 30, 32, 33 ill., 37, 49
 all-movable horizontal stabilizer, 49 n
 stability, 127
F-100A, 36, 37
Feathering a propeller, 43
Fischel, Jack, 92-108
Flights of Discovery, 59
France, 11
Fulton, Fitzhugh, 52

G force, 24, 27
Goodlin, Chalmers, 22
Gough, Ed, 22

Gough, Melvin N., 14-15, 22
Great Britain, 11
Griffith, John, x ill., xii, 12-13, 16, 20-28, 31, 50, 60

Hall, Donald, 42 ill.
Hallion, Richard P., xii, 1-2, 24, 38, 60
Hanna, Dick, 42 ill.
Hedgepeth, John T., Jr., 141
Heinemann, Edward H., 9, 11, 30, 33, 58
High-Speed Flight Research Station, 11, 15 ff., 34, 49
 radio communications, 51
 ethos, 52-53
High-Speed Flight Station, 11n, 50
High-speed instability, 53-54
Hoover, Herbert, 14-16, 15 ill., 17 ill., 18 ill.
Hyatt, Abraham, 3
Hydrogen peroxide, 30, 142
Hyland, Andrew, 39

Icing research, 23
Ikeler, Vicki, 55

Jansen, George, 32, 40 n
Jet assisted take-off, 20 n, 22 ill., 31
Jet Pilot (movie), 25-26
Jones, Robert T., 4, 28
Jones, Walter P., 40, 42 ill., 60
Jordan, Gareth H., 86-89

Kármán, Theodore von, 28
KC-135, 36
Kent, Marion I., 141
Kincaid, Gilbert W., 55-56
Kotcher, Ezra, 3

L-39, 13, 23
Landis, Tony, 11
Lane, Eddie, 51
Langley Memorial Aeronautical Laboratory, 4, 12, 14-15, 28
Lewis Laboratory, 13, 23, 25 n
Lilly, Howard C., 7, 8 ill., 15, 20, 60
 death, 7-8, 20
Liquid oxygen (lox), 55
Litchfield Park, Ariz., 45

M.52, 57
Mach 2 Dawn, 59
Mach, Ernst, xii,
Martin, John F., 1, 31, 60

May, Eugene F., 7, 8 ill., 31, 60
McKay, John B., 34-35, 40-41, 43-45, 60
Meteor, 7
Miles Aircraft Corporation, 57
Moise, John W., 55
Mojave Air Station, 65
Monocoque construction, 7
Muroc Army Air Field (later, Air Force Base—still later, Edwards AFB), xii, 17, 22 n
Muroc Dry Lake, 17 ill.
Muroc Flight Test Unit, 25 n, 46, 66
  housing, 65
Murray, Arthur, 27, 54

National Advisory Committee for Aeronautics (NACA), , 3, 7, 14, see also Langley, Lewis, Dryden
Navy, U.S., xii, 3-4, 54
  Navy technical mission to Europe, 29
North American Aviation, 37

P2B-1S, xii, 16, 29 ill., 32, 34, 36, 40-41 ill., 42 ill.-46, 44 ill., 49, 108, 119, 129-132
  Explosion of number four engine in, 44-45
P-38, xii, 3, 23
P-40 Warhawk, 23
P-51 Mustang, 23
Paperclip, Operation, 47 n
Payne, Richard E., 42 ill.
Peele, James R., 90-91
Philippine Sea, battle of, 37
Phillips, William S., 59
Pitch-up, 24-25, 31, 35, 48 ff., 98, 99
Pressure suit, 43, 51, 80, 109-115, 124, 133-135

Raczkowski, Thomas J., 39
Reaction Motors, Inc., 29, 47, 86, 143-149
Research airplanes, color for, 116
Rogers Dry Lake, 17 ill., 76, 96
Root, L. Eugene, 4, 29
Ruseckas, Joe, 51n, 113
Russell, Charles W., 42
Russell, John W., 47, 55

San Jacinto, 36
Santner, F. A., 113
Schneider, Edward T., xii, 1, 28
Sjoberg, Sigurd A., 24-25
Smith, A. M. O., 5, 29
Smith, R. G., 4, 6
Soulé, Hartley A., 64

Sound, speed of, xii
Stack, John, 3
Stalling, 48
Static stability and instability, 54 n
Sullivan, J. E., 112
Supersonic yaw, see High-speed instability
Swept wings, 4, 28-29, 48, 57
Szalai, Kenneth, 1, 58-59

TBM Avenger, 36-37
Technological revolution, 31
Tipton, Joseph L., 42 ill., 43-45
Torpedo-Bomber Air Group VT-51, 36
Transonic speed range, xi, 3, 23-24, 28, 29
Trapnell, Frederick, 60
Truax, Robert, 47
Truszynski, Gerald M., 47 n
Turndrup, Don, 40

Van Every, Kermit, 30
Vensel, Joseph, 12 ill., 25, 38-39, 42, 50
von Braun, Wernher, 47

Walker, Joseph A., 39, 40-41, 42 ill., 60, 141
Westinghouse J34 jet engine, 29-30, 52
Williams, Esther, 56
Williams, Walter C., xii, 11, 13 ill., 25, 38, 46, 116-117, 124, 126-128
Wind tunnels, 2, 28-29
Wing slats, fences, leading-edge extensions, and flaps, 30, 33, 35 ill., 48 ill.-49, 101
Woods, Robert, 4
Wright flyer, 57

X-airplanes, 58
X-1, see XS-1
X-1A, 36, 40, 54
X-1B, 40, 152
X-1D, 54
X-1E, 7, 20 ill., 40, 152
X-2, see XS-2
X-3, 37
X-4, 36, 37, 52, 123
X-5, 36, 37, 123, 127
X-15, 37, 54, 58
XF-92A, 37, 58, 123
XLR-8 (Reaction Motors 6000C4) rocket engine, 29, 47, 124-125
  modifications for Mach 2 flight, 55, 86, 136-140
  Improved version, 143-149, 151-152
XP-42, 57

XS-2, 4, 5, 13, 32, 37, 51, 54
XS-1, xii, 3-4, 7, 15 ill., 16, 17 ill., 18-19 ill., 20, 25-26, 37
  air launch techniques, 48, 49, 118-120
  all-movable horizontal stabilizer, 49 n, 56-57
  comparison of straight wing with D-558-2's swept wing, 52

Yeager, Chuck, 15, 17 ill., 27, 33, 54, 56, 119
YF-92A, misprint for XF-92A, 123

Zero-zero ejection seat (in reverse), 39
Zuikaku (Japanese carrier), 37

THE NASA HISTORY SERIES

Reference Works, NASA SP-4000:

Grimwood, James M. *Project Mercury: A Chronology*. (NASA SP-4001, 1963).

Grimwood, James M., and Hacker, Barton C., with Vorzimmer, Peter J. *Project Gemini Technology and Operations: A Chronology*. (NASA SP-4002, 1969).

Link, Mae Mills. *Space Medicine in Project Mercury*. (NASA SP-4003, 1965).

Astronautics and Aeronautics, 1963: Chronology of Science, Technology, and Policy. (NASA SP-4004, 1964).

Astronautics and Aeronautics, 1964: Chronology of Science, Technology, and Policy. (NASA SP-4005, 1965).

Astronautics and Aeronautics, 1965: Chronology of Science, Technology, and Policy. (NASA SP-4006, 1966).

Astronautics and Aeronautics, 1966: Chronology of Science, Technology, and Policy. (NASA SP-4007, 1967).

Astronautics and Aeronautics, 1967: Chronology of Science, Technology, and Policy. (NASA SP-4008, 1968).

Ertel, Ivan D., and Morse, Mary Louise. *The Apollo Spacecraft: A Chronology, Volume I, Through November 7, 1962*. (NASA SP-4009, 1969).

Morse, Mary Louise, and Bays, Jean Kernahan. *The Apollo Spacecraft: A Chronology, Volume II, November 8, 1962-September 30, 1964*. (NASA SP-4009, 1973).

Brooks, Courtney G., and Ertel, Ivan D. *The Apollo Spacecraft: A Chronology, Volume III, October 1, 1964-January 20, 1966*. (NASA SP-4009, 1973).

Ertel, Ivan D., and Newkirk, Roland W., with Brooks, Courtney G. *The Apollo Spacecraft: A Chronology, Volume IV, January 21, 1966-July 13, 1974*. (NASA SP-4009, 1978).

Astronautics and Aeronautics, 1968: Chronology of Science, Technology, and Policy. (NASA SP-4010, 1969).

Newkirk, Roland W., and Ertel, Ivan D., with Brooks, Courtney G. *Skylab: A Chronology*. (NASA SP-4011, 1977).

Van Nimmen, Jane, and Bruno, Leonard C., with Rosholt, Robert L. *NASA Historical Data Book, Volume I: NASA Resources, 1958-1968*. (NASA SP-4012, 1976, rep. ed. 1988).

Ezell, Linda Neuman. *NASA Historical Data Book, Volume II: Programs and Projects, 1958-1968*. (NASA SP-4012, 1988).

Ezell, Linda Neuman. *NASA Historical Data Book, Volume III: Programs and Projects, 1969-1978*. (NASA SP-4012, 1988).

Gawdiak, Ihor Y., with Fedor, Helen. Compilers. *NASA Historical Data Book, Volume IV: NASA Resources, 1969-1978*. (NASA SP-4012, 1994).

Astronautics and Aeronautics, 1969: Chronology of Science, Technology, and Policy. (NASA SP-4014, 1970).

Astronautics and Aeronautics, 1970: Chronology of Science, Technology, and Policy. (NASA SP-4015, 1972).

Astronautics and Aeronautics, 1971: Chronology of Science, Technology, and Policy. (NASA SP-4016, 1972).

Astronautics and Aeronautics, 1972: Chronology of Science, Technology, and Policy. (NASA SP-4017, 1974).

Astronautics and Aeronautics, 1973: Chronology of Science, Technology, and Policy. (NASA SP-4018, 1975).

Astronautics and Aeronautics, 1974: Chronology of Science, Technology, and Policy. (NASA SP-4019, 1977).

Astronautics and Aeronautics, 1975: Chronology of Science, Technology, and Policy. (NASA SP-4020, 1979).

Astronautics and Aeronautics, 1976: Chronology of Science, Technology, and Policy. (NASA SP-4021, 1984).

Astronautics and Aeronautics, 1977: Chronology of Science, Technology, and Policy. (NASA SP-4022, 1986).

Astronautics and Aeronautics, 1978: Chronology of Science, Technology, and Policy. (NASA SP-4023, 1986).

Astronautics and Aeronautics, 1979-1984: Chronology of Science, Technology, and Policy. (NASA SP-4024, 1988).

Astronautics and Aeronautics, 1985: Chronology of Science, Technology, and Policy. (NASA SP-4025, 1990).

Noordung, Hermann. *The Problem of Space Travel: The Rocket Motor.* Stuhlinger, Ernst, and Hunley, J.D., with Garland, Jennifer. Editor. (NASA SP-4026, 1995).

Astronautics and Aeronautics, 1986-1990: A Chronology. (NASA SP-4027, 1997).

Management Histories, NASA SP-4100:

Rosholt, Robert L. *An Administrative History of NASA, 1958-1963.* (NASA SP-4101, 1966).

Levine, Arnold S. *Managing NASA in the Apollo Era.* (NASA SP-4102, 1982).

Roland, Alex. *Model Research: The National Advisory Committee for Aeronautics, 1915-1958.* (NASA SP-4103, 1985).

Fries, Sylvia D. *NASA Engineers and the Age of Apollo.* (NASA SP-4104, 1992).

Glennan, T. Keith. *The Birth of NASA: The Diary of T. Keith Glennan.* Hunley, J.D. Editor. (NASA SP-4105, 1993).

Seamans, Robert C., Jr. *Aiming at Targets: The Autobiography of Robert C. Seamans, Jr.* (NASA SP-4106, 1996)

Project Histories, NASA SP-4200:

Swenson, Loyd S., Jr., Grimwood, James M., and Alexander, Charles C. *This New Ocean: A History of Project Mercury.* (NASA SP-4201, 1966).

Green, Constance McL., and Lomask, Milton. *Vanguard: A History.* (NASA SP-4202, 1970; rep. ed. Smithsonian Institution Press, 1971).

Hacker, Barton C., and Grimwood, James M. *On Shoulders of Titans: A History of Project Gemini.* (NASA SP-4203, 1977).

Benson, Charles D. and Faherty, William Barnaby. *Moonport: A History of Apollo Launch Facilities and Operations.* (NASA SP-4204, 1978).

Brooks, Courtney G., Grimwood, James M., and Swenson, Loyd S., Jr. *Chariots for Apollo: A History of Manned Lunar Spacecraft.* (NASA SP-4205, 1979).

Bilstein, Roger E. *Stages to Saturn: A Technological History of the Apollo/Saturn Launch Vehicles.* (NASA SP-4206, 1980).

SP-4207 not published.

Compton, W. David, and Benson, Charles D. *Living and Working in Space: A History of Skylab.* (NASA SP-4208, 1983).

Ezell, Edward Clinton, and Ezell, Linda Neuman. *The Partnership: A History of the Apollo-Soyuz Test Project.* (NASA SP-4209, 1978).

Hall, R. Cargill. *Lunar Impact: A History of Project Ranger.* (NASA SP-4210, 1977).

Newell, Homer E. *Beyond the Atmosphere: Early Years of Space Science.* (NASA SP-4211, 1980).

Ezell, Edward Clinton, and Ezell, Linda Neuman. *On Mars: Exploration of the Red Planet, 1958-1978.* (NASA SP-4212, 1984).

Pitts, John A. *The Human Factor: Biomedicine in the Manned Space Program to 1980.* (NASA SP-4213, 1985).

Compton, W. David. *Where No Man Has Gone Before: A History of Apollo Lunar Exploration Missions.* (NASA SP-4214, 1989).

Naugle, John E. *First Among Equals: The Selection of NASA Space Science Experiments.* (NASA SP-4215, 1991).

Wallace, Lane E. *Airborne Trailblazer: Two Decades with NASA Langley's Boeing 737 Flying Laboratory.* (NASA SP-4216, 1994).

Butrica, Andrew J. Editor. *Beyond the Ionosphere: Fifty Years of Satellite Communication.* (NASA SP-4217, 1997).

Butrica, Andrews J. *To See the Unseen: A History of Planetary Radar Astronomy*. (NASA SP-4218, 1996).

Mack, Pamela E. Editor. *From Engineering Science to Big Science: The NACA and NASA Collier Trophy Research Project Winners*. (NASA SP-4219, 1998).

Reed, R. Dale. With Lister, Darlene. *Wingless Flight: The Lifting Body Story*. (NASA SP-4220, 1997).

Center Histories, NASA SP-4300:

Rosenthal, Alfred. *Venture into Space: Early Years of Goddard Space Flight Center*. (NASA SP-4301, 1985).

Hartman, Edwin, P. *Adventures in Research: A History of Ames Research Center, 1940-1965*. (NASA SP-4302, 1970).

Hallion, Richard P. *On the Frontier: Flight Research at Dryden, 1946-1981*. (NASA SP-4303, 1984).

Muenger, Elizabeth A. *Searching the Horizon: A History of Ames Research Center, 1940-1976*. (NASA SP-4304, 1985).

Hansen, James R. *Engineer in Charge: A History of the Langley Aeronautical Laboratory, 1917-1958*. (NASA SP-4305, 1987).

Dawson, Virginia P. *Engines and Innovation: Lewis Laboratory and American Propulsion Technology*. (NASA SP-4306, 1991).

Dethloff, Henry C. *"Suddenly Tomorrow Came...": A History of the Johnson Space Center*. (NASA SP-4307, 1993).

Hansen, James R. *Spaceflight Revolution: NASA Langley Research Center from Sputnik to Apollo*. (NASA SP-4308, 1995).

Wallace, Lane E. *Flights of Discovery: 50 Years at the NASA Dryden Flight Research Center*. (NASA SP-4309, 1996).

Herring, Mack R. *Way Station to Space: A History of the John C. Stennis Space Center*. (NASA SP-4310, 1997).

Wallace, Harold D., Jr. *Wallops Station and the Creation of the American Space Program*. (NASA SP-4311, 1997).

General Histories, NASA SP-4400:

Corliss, William R. *NASA Sounding Rockets, 1958-1968: A Historical Summary*. (NASA SP-4401, 1971).

Wells, Helen T., Whiteley, Susan H., and Karegeannes, Carrie. *Origins of NASA Names*. (NASA SP-4402, 1976).

Anderson, Frank W., Jr. *Orders of Magnitude: A History of NACA and NASA, 1915-1980*. (NASA SP-4403, 1981).

Sloop, John L. *Liquid Hydrogen as a Propulsion Fuel, 1945-1959.* (NASA SP-4404, 1978).

Roland, Alex. *A Spacefaring People: Perspectives on Early Spaceflight.* (NASA SP-4405, 1985).

Bilstein, Roger E. *Orders of Magnitude: A History of the NACA and NASA, 1915-1990.* (NASA SP-4406, 1989).

Logsdon, John M. Editor. With Lear, Linda J., Warren-Findley, Jannelle, Williamson, Ray A., and Day, Dwayne A. *Exploring the Unknown: Selected Documents in the History of the U.S. Civil Space Program, Volume I, Organizing for Exploration.* (NASA SP-4407, 1995).

Logsdon, John M. Editor. With Day, Dwayne A., and Launius, Roger D. *Exploring the Unknown: Selected Documents in the History of the U.S. Civil Space Program, Volume II, Relations with Other Organizations.* (NASA SP-4407, 1996).

Logsdon, John M. Editor. With Launius, Roger D., Onkst, David H., and Garber, Stephen. *Exploring the Unknown: Selected Documents in the History of the U.S. Civil Space Program, Volume III, Using Space.* (NASA SP-4407, 1998).

www.ingramcontent.com/pod-product-compliance
Lightning Source LLC
Chambersburg PA
CBHW080545170426
43195CB00016B/2686